T0321043

Gene-Environment
Interaction Analysis

Gene–Environment Interaction Analysis

Methods in Bioinformatics and Computational Biology

edited by

Sumiko Anno

PAN STANFORD PUBLISHING

Published by

Pan Stanford Publishing Pte. Ltd.
Penthouse Level, Suntec Tower 3
8 Temasek Boulevard
Singapore 038988

Email: editorial@panstanford.com
Web: www.panstanford.com

British Library Cataloguing-in-Publication Data
A catalogue record for this book is available from the British Library.

Gene–Environment Interaction Analysis:
Methods in Bioinformatics and Computational Biology

ISBN   978-981-4669-63-4 (Hardcover)
ISBN   978-981-4669-64-1 (eBook)

Printed in the USA

Contents

Preface

Gene–environment (G × E) interactions contribute to the development of complex diseases and phenotypic variation. They are a hot topic in human genetics, and analyses of G × E interactions are expected to have many potential applications. Despite the importance of G × E interactions in the etiology of complex diseases and phenotypic variation, insufficient attention has been paid to developing models for detecting these interactions. This textbook introduces different models of G × E interactions for use in the determination of human disease and phenotypic variation. Applying models of the complex interactions between genes and environment will lead to novel methods of disease detection and prevention, as well as new interventions in various fields.

I would like to thank the publisher for bringing me along on this adventure and for doing an excellent job. I am particularly grateful to Stanford Chong, Sarabjeet Garcha, and Shivani Sharma.

Sumiko Anno
February 2016

Chapter 1

Understanding Skin Color Variations as an Adaptation by Detecting Gene–Environment Interactions

Sumiko Anno,[a] Kazuhiko Ohshima,[b] and Takashi Abe[c]

[a]*Shibaura Institute of Technology, 3-7-5, Toyosu, Koto-ku, Tokyo 135-8548, Japan*
[b]*Nagahama Institute of Bio-Science and Technology, 1266 Tamura-cho, Nagahama-shi, Shiga 526-0829, Japan*
[c]*Niigata University, 8050, Ikarashi 2-no-cho, Nishi-ku, Niigata 950-2181, Japan*
takaabe@ie.niigata-u.ac.jp, annou@sic.shibaura-it.ac.jp,
k_ohshima@nagahama-i-bio.ac.jp

Genetic and environmental factors influence the elaborate feedback mechanism that enables the human adaptive form to make internal adjustments in response to environmental stimuli. Human survival may ultimately depend on research elucidating the complex dynamics of the human genome, as well as an understanding of how environmental pressures affect the genome and influence human traits. This chapter reviews our present knowledge of the mechanisms by which haplotypes comprising multiple single-nucleotide polymorphisms (SNPs) can contribute to differences between human popu-

Gene–Environment Interaction Analysis: Methods in Bioinformatics and Computational Biology
Edited by Sumiko Anno

lation groups. Herein, we describe current approaches to detecting natural selection in pigmentation candidate genes on the basis of haplotypes revealed by SNP analyses. This chapter also discusses methods for elucidating the selective genetic mechanisms that have operated to alter human skin pigmentation, which may be induced by ultraviolet radiation (UVR) in the birthplaces of human populations. Finally, we present our recommendation of spatial statistical methods for clarifying gene–environment interactions, as applicable to interactions with UVR levels. Spatial statistical approaches that apply environmental association rules can be used to extend our knowledge of human adaptation to the environment.

1.1 Introduction

The increasing pace of industrial technological advancement has contributed to environmental deterioration. Mass consumption of resources and energy, widespread expansion of economic activities, alteration of the environment by large-scale development, etc., have caused global environment changes, such as global warming and ozone layer depletion. These changes may threaten not only the environment but also human health. For example, industrial activity has sped up the rate of ozone depletion, leading to increased exposure to ultraviolet (UV) rays—a condition that can cause human health problems.

Overcoming these new and rapid global environment changes may require that we learn the true characteristics of environmental adaptation and improve the breadth of these abilities. Genetic and environmental factors influence the elaborate feedback mechanism that enables the human adaptive form to make internal adjustments in response to environmental stimuli [1]. Human survival could ultimately depend on our understanding of two important adaptation components. First, we have to understand the complex dynamics of the human genome. Second, we must further investigate how the environment exerts pressures that can affect the genome in a manner that determines human traits.

1.1.1   Human Skin Color as an Environmental Adaptation

Humans can adapt to complicated and challenging environments by evolving new traits and abilities. Human skin color variations exist due to increase and decrease in melanin, which represent environmental adaptations to different levels of exposure to ultraviolet radiation (UVR) [2, 3]. For example, people indigenous to Africa have dark skin because melanin production is increased at low latitudes to protect against repeated UV irradiation. In contrast, people indigenous to northern Europe have pale skin because melanin production is decreased at high latitudes, which increases the body's ability to synthesize vitamin D and offers a variety of health benefits, including protection against rickets (osteomalacia).

1.1.2   Mechanism of Melanin Formation

Human cutaneous pigmentation is primarily determined by melanin—not by the number of melanocytes but by the size, shape, distribution, and chemical composition of black/brown eumelanin and yellow/red pheomelanin. Both types of melanin are synthesized in the melanosome (an organelle contained within melanocytes). Melanin formation begins when UV irradiation of human melanocytes triggers the binding of alpha-melanocyte-stimulating hormone (α-MSH) to the G-protein-coupled melanocortin 1 receptor (*MC1R*), activating adenylate cyclase and increasing cyclic adenosine monophosphate (cAMP) formation. The increased intracellular cAMP ultimately stimulates the expressions of enzymes that are important in eumelanin biosynthesis—including tyrosinase, which catalyzes the oxidation of tyrosine to dopaquinone, the last common precursor of eumelanin and pheomelanin. The fate of dopaquinone is largely determined by the signaling state of the *MC1R*; the binding of agouti to *MC1R* activates the signal transduction events that lead to pheomelanin synthesis [4, 5]. Differences in human pigmentation are understood to largely result from differences in the eumelanin content, which is determined by melanin productivity and is genetically controlled [4, 5].

1.1.3 Skin Color Diversity due to DNA Polymorphism

Although approximately 99% of human DNA sequences are reportedly identical across the human population, variations in the DNA sequence exert an important impact on human diversity, affecting disease risk, drug responsiveness (including adverse drug responses), and human phenotypes [4, 5]. The ability to determine the functional influence of genetic variations could offer the possibilities of individualized treatment, disease prevention, and elucidation of the evolutionary and biophysical functions that characterize human phenotypes. A single-nucleotide polymorphism (SNP) is a genetic variation that occurs in at least 1% of the population. Studies of skin pigmentation have reported skin color–associated polymorphisms at several loci, including *MC1R*, oculocutaneous albinism II (P), and agouti-signaling protein (ASIP). It is thought that human skin color variation is likely controlled by interactions between SNP alleles at multiple loci [3–10].

1.1.4 Objectives

The ozone layer prevents the most harmful UVB wavelengths from passing through the earth's atmosphere; therefore, ozone depletion by chlorofluorocarbons (CFCs) is an important environmental issue. Increased UV exposure is thought to have a variety of biological consequences, including an increased incidence of skin cancer, damage to plants, and reduction of plankton populations in the ocean photic zone. Human skin color variations are known to occur in connection with environmental factors, such as UVR; however, the molecular basis for the genetic background of human skin color remains unclear. Further studies are needed to enable predictions of how environmental changes—such as increased UV rays due to ozone depletion—might influence human health in the future and to determine how humans can better adapt to the changing ecosystem.

In this chapter, we review the presently available information regarding the mechanisms of haplotype formation from SNP variations, a process that contributes to substantial differences between population groups. We discuss current approaches that use SNP analysis to detect natural selection of pigmentation candidate genes, as well as approaches to elucidate UVR-induced selective genetic mechanisms of darkening and lightening human skin color. Herein, we also describe spatial statistical methods for clarifying gene–environment interactions, which are applicable to interactions with UVR levels, as well as the technologies of genetic engineering, remote sensing (RS), and geographic information systems (GIS). We ultimately recommend a spatial statistical approach that applies environmental association rules and can thus extend our knowledge of human adaptation to the environment.

1.2 A Linkage Disequilibrium–Based Statistical Approach to Detect Interactions between SNP Alleles at Multiple Loci That Contribute to Skin Pigmentation Variation between Human Populations

1.2.1 SNP Analysis of European and East Asian Cohorts

Our group performed a study comparing European and East Asian cohorts with different skin colors. Genetic data were obtained from 122 Europeans subjects in the U.S. and 100 East Asian subjects in Japan. From all subjects, buccal samples were collected and anonymously coded, and the melanin skin pigmentation index was determined using the Mobile Mexameter MSC100/MX18 (Integral Corporation, Shinjuku-ku, Tokyo, Japan), which uses photodiode arrays to measure the intensity of particular wavelengths of light. Two measurements were obtained from each subject's back as an index of inherent skin color influenced by genetic factors, and two measurements were obtained from the cheek as an index of modified skin color influenced by environmental factors.

DNA was extracted from the buccal samples, and to ensure sufficient genomic DNA for SNP genotyping, whole genomic DNA was amplified using the REPLI-g Kit (QIAGEN K. K., Chuo-ku, Tokyo, Japan). Candidate genes for human skin pigmentation were identified [11–23], including *ASIP*, tyrosinase-related protein 1 (*TYRP1*), tyrosinase (*TYR*), *MC1R*, oculocutaneous albinism II (*OCA2*), microphthalmia-associated transcription factor (*MITF*), and myosin VA (*MYO5A*).

From loci in these candidate genes, we selected 20 SNPs that had been registered in the single nucleotide polymorphism database (dbSNP) [24]: rs819136, rs1129414, rs2075508, rs10960756, rs3793976, rs2298458, rs3212363, rs1805008, rs3212371, rs2279727, rs4778182, rs1800419, rs2311843, rs1800414, rs1800404, rs7623610, rs704246, rs16964944, rs1724577, and rs4776053.

The allele discrimination assay consisted of polymerase chain reaction (PCR) amplification of multiple SNP alleles at a particular locus using specific primers with tags differing in molecular weight. Purified PCR products were combined with two heminested allele-specific primers and two universally tagged Masscode oligonucleotide primers. Each tag was covalently attached to the 5' end of an oligonucleotide primer via a photolabile linker. Following PCR amplification, the SNP-specific PCR products were passed through a QIAquick 96 silica-based filter membrane to remove unincorporated tagged primers. The filtered PCR products were exposed to a 254 nm mercury lamp to cleave the incorporated tags, and the tags were analyzed using an optimized Agilent 1100 single-quadrupole mass spectrometer. The presence of a particular tag indicated the presence of the corresponding SNP allele in the genomic DNA sample. Genotype data were reported in a comma-delimited flat-file format that contained the SNP and sample identifiers for each detected allele. Alleles were reported using binary nomenclature, in which 1 represented wild-type alleles and 2 represented variant alleles. The SNP allele was classified as one of three types: wild-type homozygous (1,1), variant-type homozygous (2,2), and heterozygous (1,2) [25, 26].

1.2.2 Genotype and Allele Frequencies for the 20 SNPs in the European and East Asian Population Groups

We next determined genotype and allele frequencies in the European and East Asian cohorts in order to determine race-related differences in their distributions. Table 1.1 and Figs. 1.1 and 1.2 show the genotype and allele frequencies for the 20 SNPs that were speculated to contribute to skin color differences between the two population groups.

Table 1.1 Genotypes and allele frequencies for 20 SNPs in European and East Asian populations

rs #	rs819136		rs1129414		rs2075508		rs10960756		rs3793976	
Allele type	A/G		A/C		C/T		A/G		G/T	
Population	European	East Asian	European	East Asian	European	East Asian	European	East Asian	European	East Asian
Genotypes 1/1	0.008	0.050	1.000	1.000	0.000	0.010	0.000	0.130	0.884	0.750
Genotypes 1/2	0.180	0.280	0.000	0.000	0.180	0.380	0.008	0.460	0.116	0.230
Genotypes 2/2	0.811	0.670	0.000	0.000	0.820	0.610	0.992	0.410	0.000	0.020
Alleles 1	0.098	0.190	1.000	1.000	0.090	0.200	0.004	0.360	0.942	0.865
Alleles 2	0.902	0.810	0.000	0.000	0.910	0.800	0.996	0.640	0.058	0.135

rs #	rs2298458		rs3212363		rs1805008		rs3212371		rs2279727	
Allele type	G/T		A/T		C/T		A/G		A/C	
Population	European	East Asian	European	East Asian	European	East Asian	European	East Asian	European	East Asian
Genotypes 1/1	0.943	0.820	0.525	0.010	0.885	1.000	0.803	0.780	0.918	0.040
Genotypes 1/2	0.057	0.160	0.443	0.384	0.115	0.000	0.197	0.210	0.082	0.420
Genotypes 2/2	0.000	0.020	0.033	0.606	0.000	0.000	0.000	0.010	0.000	0.540
Alleles 1	0.971	0.900	0.746	0.202	0.943	1.000	0.902	0.885	0.959	0.250
Alleles 2	0.029	0.100	0.254	0.798	0.057	0.000	0.098	0.115	0.041	0.750

rs #	rs4778182		rs1800419		rs2311843		rs1800414		rs1800404	
Allele type	A/G		C/T		C/T		A/G		A/G	
Population	European	East Asian	European	East Asian	European	East Asian	European	East Asian	European	East Asian
Genotypes 1/1	0.156	0.000	0.142	0.180	0.725	0.040	1.000	0.150	0.615	0.140
Genotypes 1/2	0.533	0.000	0.492	0.450	0.258	0.270	0.000	0.440	0.311	0.440
Genotypes 2/2	0.311	1.000	0.367	0.370	0.017	0.690	0.000	0.410	0.074	0.420
Alleles 1	0.422	0.000	0.388	0.405	0.854	0.175	1.000	0.370	0.770	0.360
Alleles 2	0.578	1.000	0.613	0.595	0.146	0.825	0.000	0.630	0.230	0.640

rs #	rs7623610		rs704246		rs16964944		rs1724577		rs4776053	
Allele type	A/G		C/T		A/G		G/T		C/T	
Population	European	East Asian	European	East Asian	European	East Asian	European	East Asian	European	East Asian
Genotypes 1/1	0.302	0.170	0.615	0.450	1.000	0.950	0.992	0.950	0.689	0.580
Genotypes 1/2	0.388	0.440	0.320	0.440	0.000	0.050	0.008	0.050	0.287	0.360
Genotypes 2/2	0.310	0.390	0.066	0.110	0.000	0.000	0.000	0.000	0.025	0.060
Alleles 1	0.496	0.390	0.775	0.670	1.000	0.975	0.996	0.975	0.832	0.760
Alleles 2	0.504	0.610	0.225	0.330	0.000	0.025	0.004	0.025	0.168	0.240

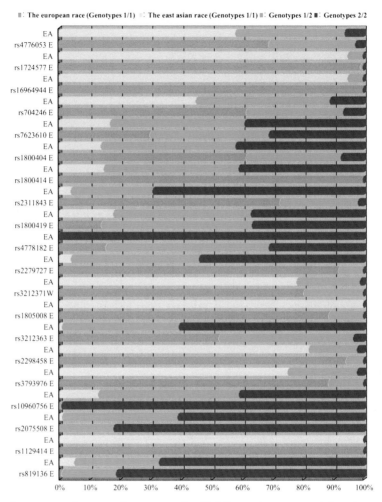

Figure 1.1   Genotype frequencies for 20 SNPs in European and East Asian populations. E: European; EA: East Asian.

1.2.3   Cluster Analysis

We next performed cluster analysis for genetic differentiation of the SNP genotyping results. We condensed the genotype assignment for each SNP allele into a single numeric value, as follows: homozygous wild type $1,1 = 0$; heterozygous $1,2 = 0.5$; and homozygous variant $2,2 = 1$. The genotyping data were then used to construct an unweighted pair group method with arithmetic mean (UPGMA) dendrogram using the Euclidean distance. UPGMA is one of the simplest and most commonly used hierarchical clustering algorithms; it constructs a hierarchical dendrogram based on the input of a set of components and a distance matrix containing pairwise distances between all components. Cluster analysis showed that each racial group formed a separate cluster, except for one East Asian subject who was included in the European cluster (Figs. 1.3 and 1.4).

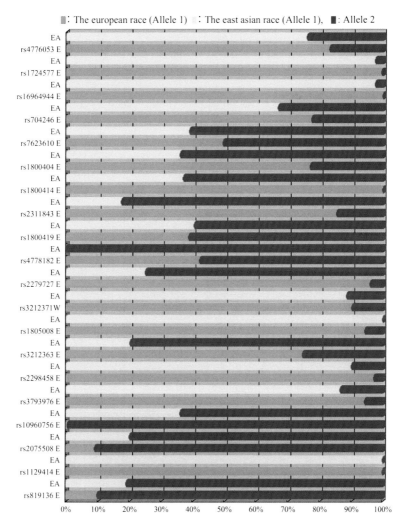

Figure 1.2   Allele frequencies for 20 SNPs in European and East Asian populations. E: European; EA: East Asian.

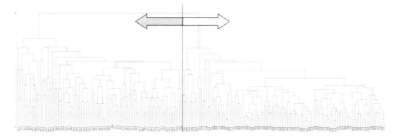

Figure 1.3   Dendrogram obtained from cluster analysis. Racial groups are separated by a red line. Left: East Asian subjects; right: European subjects.

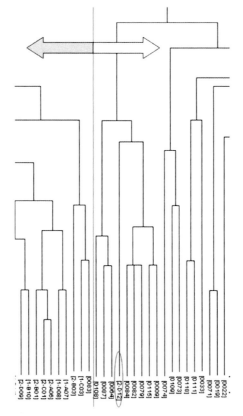

Figure 1.4   An enlarged view of a part of the dendrogram. Racial groups are separated by a red line, with East Asian subjects to the left and European subjects to the right, except for one East Asian subject circled in red.

1.2.4   Linkage Disequilibrium Generated by Gene–Gene Interactions Contributes to Differences between Racial Groups

To examine the contributions to racial differences of nonrandom associations of SNP alleles at multiple loci, we examined the associations between the 20 SNP alleles at various candidate gene loci in the genome by calculating linkage disequilibrium (LD). LD is the association between the qualitative random variables corresponding to SNP alleles at different polymorphic sites that are not necessarily on the same chromosome [27, 28] and serves as a measure of gene–gene interactions among unlinked loci [29]. LD provides important gene-mapping information when used as a tool for fine mapping of complex disease genes and in genome-wide association studies. LD may also reveal information about the evolution of populations.

In the described study, we used the concept of LD that was defined by Richard Lewontin in one of the earliest proposed measures of disequilibrium (symbolized by D) [30]. When measuring LD, D quantifies disequilibrium as the difference between the observed frequency of a two-locus haplotype and the frequency that would be expected if the alleles segregated at random. The haplotype frequency of two markers with alleles A and a and B and b can be described as f_{AB}, f_{Ab}, f_{aB}, and f_{ab}, and the discrepancy of the distribution under LD can be measured by

$$D = f_{AB}f_{ab} - f_{Ab}f_{aB} \tag{1.1}$$

Measures of LD are defined as the standardized values of D. Two common such measures are

$$R^2 = D^2/(f_{AB} + f_{Ab})(f_{AB} + f_{aB})(f_{aB} + f_{ab})(f_{Ab} + f_{ab})$$

and $D' = D/D_{max}$,

where when the numerator is positive D_{max} is

$$\min([f_{AB} + f_{Ab}][f_{Ab} + f_{ab}], [f_{AB} + f_{aB}][f_{aB} + f_{ab}])$$

and otherwise

$$\min([f_{AB} + f_{Ab}][f_{AB} + f_{aB}], [f_{Ab} + f_{ab}][f_{aB} + f_{ab}]) \tag{1.2}$$

The case where $D' = 1$ is known as complete LD. D' values of <1 indicate that the complete ancestral LD has been disrupted and there is no clear interpretation of D' values of <1. The definition of R^2 can be understood by considering the alleles as realizations of quantitative random variables with values of 0 and 1, from which a correlation coefficient can be calculated. LD can be analyzed with software such as the estimate haplotype frequencies (EH) program [31], Haploview [32], R statistical software [33], and others.

We calculated the LD statistic and significance levels for all possible SNP allele pairs. Significance levels were calculated using a χ^2 test on the 2 × 2 table of haplotype frequencies. The p value of LD was determined with a χ^2 test; statistical significance was set at 0.05. Combinations of SNP alleles at multiple loci under LD were jointly tested for association with the European or East Asian group by performing a χ^2 test for independence. Only data that fit the Hardy–Weinberg equilibrium were used in the analysis.

We found the allele combination rs1800419-C/rs1800414-G/rs1800404-G to be associated with the East Asian group (p = 5.39 × 10^{-20}). These alleles are found in *OCA2* on chromosome 15 and form a haplotype [34]. *OCA2* controls the transport of the melanin precursor tyrosine into the melanosome for melanin synthesis.

The allele combination rs2311843-C/rs1800404-A/rs4776053-C was associated with the European group (p = 5.51 × 10^{-33}). These alleles are found in *OCA2* and *MYO5A* on chromosome 15 and form a haplotype [34]. *MYO5A* functions in vesicle transport, and mutations in this gene confer a lighter skin color due to defects in actin-based pigment granule transport within melanocytes. This pigmentation variation is believed to be due to abnormal distribution of melanosomes along melanocyte dendrites [35].

There were significant differences in the allele combinations (i.e., haplotypes) between the two racial groups. rs4776053 in *MYO5A* was found only in the European group, while rs2311843 and rs1800404 in *OCA2* were found in both the East Asian and European groups. Thus, only the rs4776053 allele associated with the European group could be considered to confer lighter skin color. The rs4776053 allele exists as C or T, with allele frequencies of 83.2/16.8 (C/T) for the European group and 76.0/24.0 (C/T) for the East Asian group. The higher frequency of the C variant within the European group indicates that rs4776053-C could be a SNP allele that confers lighter skin color. This finding suggests that the lighter skin pigmentation observed in European populations results from the positive selection of different loci in different human populations [11, 34].

1.3 SNP Analyses Reveal Natural Selection of Pigmentation Candidate Genes from Haplotypes

1.3.1 Background

In recent years, attempts have been made to detect recent natural selection in the human genome on the basis of the haplotype structure. The ability to detect recent natural selection in humans would have profound implications in the fields of human adaptation/ evolutionary history and medicine [36]. Despite many studies of the biochemistry and cytology of pigmentation, few have investigated the evolutionary genetic history of this phenotypic trait [37]. A number of hypotheses involving genetic adaptation have been proposed to explain human skin color variations, but there remains only limited genetic evidence of positive selection [38].

As described above, we have studied skin pigmentation in a European cohort from the U.S. and an East Asian cohort from Japan [34]. To examine the contribution to racial differences of nonrandom associations of SNP alleles at multiple loci, we genotyped 20 SNPs in seven candidate genes and used these data to analyze the associations of the 20 SNP alleles by LD. We found that SNP alleles at multiple loci can be considered a haplotype, contributing to significant differences between two populations, and our results suggest a high probability of LD.

From these data, we selected the core haplotypes that appeared to be candidate pigmentation genes with signatures of natural selection and environmental adaptability. We conducted further analyses to find evidence of natural selection in these genes in order to shed light on the genetic evolutionary history of human skin pigmentation.

1.3.2 Detecting Natural Selection in the Human Genome on the Basis of the Haplotype Structure

We studied the rs2311843/rs1800404/rs4776053 haplotype located within *OCA2* and *MYO5A* on chromosome 15 in the European population group and the rs1800419/rs1800414/rs1800404 haplotype located within *OCA2* on chromosome 15 in the East Asian population group. Using DNA sequence polymorphism (DnaSP) version 5, we conducted both Tajima's D test and Fu and Li's F test to determine whether any of the SNPs were under natural selection pressures [39]. Only data that fit the Hardy–Weinberg equilibrium were used in the analysis. We also tested for correlations between the haplotypes and the melanin content. As described in Section 1.2.1, data on melanin content were obtained from melanin index measurements [34].

Tajima's D value of the rs1800419/rs1800414/rs1800404 haplotype in the East Asian population was significantly positive (D = 2.83967, $p < 0.01$), suggesting that this haplotype located within *OCA2* in the East Asian population has been under selective pressure. However, Fu and Li's F test showed a nonsignificant value for this haplotype (F = 1.70954, $0.05 < p < 0.10$). Significant indications of natural selection pressure were not found for any other haplotypes.

Because Tajima's D value of the rs1800419/rs1800414/rs1800404 haplotype in the East Asian population was significantly positive, we tested for correlations between this haplotype and the melanin content in this population. The frequency of rs1800419-(C/T)/rs1800414-(A/G)/rs1800404-(A/G) in the East Asian population was estimated using the EH program. For the association study, we used three haplotypes with high frequencies: rs1800419-C/rs1800414-G/rs1800404-G (0.30), rs1800419-T/rs1800414-A/rs1800404-A (0.25), and rs1800419-T/rs1800414-G/rs1800404-G (0.29). The mean melanin values in the East Asian subjects with homozygous haplotypes were C/G/G, 129.09 (n = 9); T/A/A, 162.53 (n = 8); and T/G/G, 194.07 (n = 9). In the East Asian subjects with heterozygous haplotypes C/G/G and T/G/G the mean melanin value was 144.81 (n = 35). The number of individuals with heterozygous haplotypes was greater than the expected number of 25 estimated by the haplotype frequency.

1.3.3 Discussion

These findings indicate the presence of selective pressure on *OCA2* in the East Asian population. In general, a significantly positive Tajima's *D* value is considered to be caused by the presence of balancing selection that maintains a mutation at middle frequency or a reduction of population size. Because the East Asian cohort in our study was a random sample of subjects from various regions in Japan, the influence of population size cannot be considered. Therefore, the variations in *OCA2* in the East Asian population can be considered to have been maintained by balancing selection. We also observed differences in the melanin content among the East Asian subjects with the three haplotypes (homozygous) that occurred at high frequencies.

Edwards et al. [40] recently showed that the nonsynonymous polymorphism rs1800414 (His615Arg) located within *OCA2* is significantly associated with skin pigmentation in a sample of individuals of East Asian ancestry living in Canada. We found that the number of subjects with heterozygous haplotypes containing rs1800414 was larger than the expected value. This finding provides us with new insights into the genetic evolutionary history of human skin pigmentation/human adaptation to local environments in the East Asian population.

The observed differences in mean melanin values among the haplotypes suggest that the haplotype diversity in *OCA2* may play an important role in maintaining the diversity of melanin content. This case may be similar to that of sickle-cell disease (SCD); the sickle gene has a genetic advantage, known as the heterozygote advantage, that protects heterozygous carriers from succumbing to endemic *Plasmodium falciparum* malaria infection. With the increased premature death rate of homozygous individuals, the sickle gene is an example of a balanced polymorphism. In the case of pigmentation, variations in pigmentation candidate genes and melanin content in the East Asian population may similarly be considered to have been maintained by balancing selection. Evidence of balancing selection in the *OCA2* region implies that the haplotype may be associated with melanin content. This physiological trait may be related to another function of *OCA2* or a region in tight LD with the gene. Because human pigmentation is a result of the expression of multiple genes, this hypothesis requires further analysis.

1.3.4 Future Investigations of Natural Selection and Environmental Adaptability Regarding Skin Pigmentation

Various tests for detecting natural selection are expected to lead to the discovery of genes associated with environmental adaptability. Regions of the human genome undergoing positive selection have been examined using SNP databases derived from different population groups; such studies have employed a variety of genetic tests to scan for signatures, including decreased heterozygosity (the natural log of the ratio of heterozygosity [lnRH]), atypical levels of population differentiation of alleles (the fixation index [F_{st}]), and decay of LD (extended haplotype homozygosity [EHH]). A combination of these approaches has uncovered clear evidence of the selection of pigmentation-related genes in different populations [41–44]. Future investigations will require the development of new integrated tests for detecting natural selection. The analysis of functional candidate genes subject to natural selection might identify new common susceptibility (or protective) alleles for future research.

Our findings demonstrate that *OCA2* carries a strong signature of balancing selection in the East Asian population, provide new insights into the genetic evolutionary history of human skin pigmentation, and advance our knowledge about the history of human adaptation to local environments. The identification of more candidate genes involved in skin pigmentation and future analyses and clarifications of functional characteristics regarding the mutation of alleles located within candidate loci will help further elucidate the adaptive nature of human skin color. Furthermore, measuring the relationships between genetic and environmental factors will provide a clearer understanding of environmental adaptability.

The next step will be to clarify how many genes are involved, whether and how alleles interact with one another, and whether and how environmental and cultural factors influence natural selection. Further elucidation of the relationships between molecular genetics, environmental and cultural factors, and pathogenesis/phenotypic variations will facilitate the development of effective preventive, diagnostic, and therapeutic interventions and elucidate the evolutionary history.

1.4 Elucidation of the UVR-Induced Selective Genetic Mechanisms Influencing Variations in Human Skin Pigmentation

1.4.1 Influence of UVR Levels on Skin Color Variation

A number of authors have hypothesized that natural selection influences human skin color [11–13, 37, 38,41, 43, 45]. Supporting this hypothesis, recent quantitative data have indicated strong correlations between skin reflectance and UVR levels. The skin color distributions in indigenous populations have been quantitatively related to remotely sensed UVR measurements, demonstrating a strong correlation between skin reflectance and absolute latitude/UVR levels [46]. This report further suggests that global variation in skin pigmentation may result from localized adaptation to different UVR conditions via natural selection [46, 47]. A more recent study demonstrated that human skin pigmentation is the product of two clines produced by natural selection, which adjust the levels of constitutive pigmentation based on UVR levels. One cline was generated by high UVR near the equator and led to the evolution of dark, photoprotective, eumelanin-rich pigmentation. The other cline was produced by the requirement for UVB photons to sustain cutaneous photosynthesis of vitamin D_3 in low-UVB environments, resulting in the evolution of depigmented skin [48].

The evolution of depigmented skin at high latitudes has long been known to be related to vitamin D production in the skin under conditions of reduced sunlight [49, 50]. Vitamin D_3 is made in the skin when UVR penetrates the skin and is absorbed by 7-dehydrocholesterol (7-DHC) in the epidermis and dermis to form previtamin D_3. This reaction only occurs in the presence of wavelengths of 290–310 nm in the UVB range, with peak conversion occurring at 295–297 nm [48]. UVR in the range of 260–320 nm is most absorbed by nucleotides and frequently leads to deamination [51]. There was apparently strong natural selection promoting continued vitamin D production through the loss of constitutive pigmentation under conditions of reduced UVR, and the independent action of this selection on hominin populations dispersing to low-UVR habitats was inferred before there was genetic evidence demonstrating positive selection for depigmentation [46]. These conclusions suggest that skin color has likely been a selectable trait in ancestral populations at diverse geographical sites around the world [52].

1.4.2 Statistical Methods of Clarifying Gene–Environment Interactions, Viewing Skin Pigmentation as a Complex Trait

By definition, a phenotype is the result of interactions between genes and the environment [53]. Elucidation of the relationships between genetic polymorphisms and environmental exposures provides insights into pathway mechanisms for complex traits [54]. Understanding the role of adaptive evolution in skin pigmentation will require gene–environment interaction analysis approaches that integrate the investigations of genetic and environmental factors into a coherent biological framework. Fortunately, several methods have been developed for detecting gene–environment interactions for complex traits.

Murcray et al. presented a two-step approach to evaluate multiplicative gene–environment interactions in the context of gene–environment interaction studies [54]. On the basis of genome-wide data, their procedure complements screening for marginal genetic effects and thus may uncover novel genetic signals [54].

Duell et al. applied data-mining and data-reduction methods to detect interactions in epidemiological data. Multifactor dimensionality reduction, focused interaction testing framework software programs, and logistic regression models were used to evaluate pathway-based gene–gene and gene–environment interactions in a complex disease using data from a population-based, case–control study [55].

McEachin et al. used an integrated bioinformatics approach, mining multiple databases to develop and refine a model of gene-by-environment interaction that is consistent with the comorbidity of depression and alcoholism. The validity of a genetic model was established by querying publicly available databases and identifying and validating tumor necrosis factor (*TNF*) and methylenetetrahydrofolate reductase (*MTHFR*) as candidate genes. Disease genes were prioritized through analysis of common elements, showing that *TNF* and *MTHFR* shared significant commonality, consistent with a response to environmental exposure to ethanol. Finally, the MetaCore analysis program was used to model a gene-by-environment interaction that was consistent with the data [56].

In summary, studies detecting gene–environment interactions in complex traits and diseases have highlighted several successful approaches. These studies have identified high-risk subgroups in a population [54], found novel potential risk–factor combinations [55], provided insight into pathway mechanisms for complex diseases [54], and maximized the use of available biomedical data to improve understanding of complex disease [56]. These methods have been successfully applied to common complex traits believed to result from the combined effects of genes and environmental factors, and their interactions; therefore, they are applicable to complex human phenotypes, including skin color variation.

1.4.3 New Methods for Detecting Gene–Environment Interactions for Human Skin Pigmentation Variation

There has been little progress in the development of methods specifically to detect the interactions involved in skin color variation. To investigate the evolution and adaptation mechanisms involved in skin color variation, we recommend a spatial data-mining approach that provides a useful and powerful theoretical framework for detecting gene–environment interactions using the technologies of genetic engineering, RS, and GIS (Fig. 1.5). This approach can reveal environment-based differences in genetic traits and maximize the use of available data. Spatial data mining can identify biological mechanisms that regulate human adaptability to UV rays. Identifying the complex mechanism that has shaped skin color variation may advance our understanding of the history of human adaptation to local environments, as well as have important implications for public health [52].

Our proposed framework includes genetic engineering as a molecular biological approach and RS/GIS as an ecological approach in order to elucidate the relationships between genetic characteristic and UVR exposure. This method provides insights into the mechanisms for skin color variation by detecting gene–environment interactions for this complex trait. As described above, we have shown that a haplotype comprising SNP alleles at multiple loci can contribute to significant differences between the two population groups and that natural selection has likely occurred in pigmentation candidate genes from this haplotype. These findings suggest that global skin pigment variation may result from localized adaptation to different UVR conditions via natural selection [46–48]. To combine this information with an ecological approach, we used UVR information from RS data collected by a sensor that is mounted on a satellite and integrated it into GIS for spatial analysis. We then investigated the relationships between genetic characteristics and UVR exposure in order to further elucidate the role of adaptive evolution of skin pigmentation in response to environmental influences.

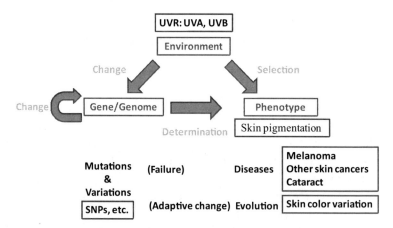

Figure 1.5 A conceptual framework of the two approaches to the gene–environment interactions for skin color variation.

1.4.4 RS Data Processing and Spatial Analysis in GIS

The UVR data were derived from readings taken from the NASA Total Ozone Mapping Spectrometer (TOMS), which was flown aboard the Nimbus-7/Earth Probe satellites. The TOMS sampled single wavelengths representative of long-wave and medium-wave UVR: 324 nm and 380 nm for UVA (range 315–400 nm) and 305 nm and 310 nm for UVB (range 280–315 nm). The original data set was very large, comprising over 64,800 readings taken each day from 1979 to 2003. Abridged data sets were produced for each wavelength, taking the average of each month from 1979 to 2003. The abridged data sets for 310 nm UVB (which induces deamination that causes a barely perceptible reddening of light skin) were then integrated with the point of longitude and latitude into a GIS for spatial analysis and interpolated using inverse distance weighting (IDW) for the values of a raster. The raster layer was overlaid with the polygon layer, which contained three polygons representing the birthplaces of the human race (Africans, Europeans, and East Asians). Each polygon contained and summarized the raster values within its area and reported the results to a table for spatial statistical analysis. Chaplin's study [47] found that the evolution of skin reflectance could be almost fully modeled as a linear effect of UVR in autumn alone. The monthly mean values by the human racial population were organized for the seasonal mean values for winter, spring, summer, and autumn.

1.4.5 Gene–Environment Interaction Analysis

With the data obtained from these analyses using genetic engineering, RS, and GIS, we then used spatial statistical analysis to evaluate gene–environmental interaction. We assessed the relationship between the SNPs at candidate genes for human skin pigmentation and the seasonal UVR exposure in the three polygon regions. This analysis was intended to clarify the mechanism of the molecular basis for the genetic background of human skin color variation, as well as human adaptability from the perspective of human evolution.

To perform this analysis, we extracted the data for the 20 SNPs in 7 candidate genes for human skin pigmentation that were obtained from our previous study, as well as for 553 SNPs in the *OCA2* gene (a common gene found in both European and East Asian populations in our previous study). We then calculated the mean values for the heterozygous genotype frequency for every allele pattern in each population.

Next, we conducted principal component analysis (PCA) using the correlation matrix in order to determine the relationships between SNP frequencies for a heterozygous subject and the seasonal UVR data. The following four kinds of PCA were conducted: (I) 20 SNPs in 7 candidate genes from our previous study + UVR data for the autumn; (II) 20 SNPs in 7 candidate genes from our previous study + UVR data for all four seasons; (III) 553 SNPs in the *OCA2* gene from the HapMap database + UVR data for the autumn; and (IV) 553 SNPs in the *OCA2* gene from the HapMap database + UVR data for all four seasons.

PCA is a statistical method that uses an orthogonal transformation to convert a set of observations of possibly correlated variables into a set of linearly uncorrelated variables called principal components [57, 58]. This method has been applied in bioinformatics for multivariate data obtained from microarrays analyses and comparative genomic analysis [59, 60]. Here, the *k*th principal component is expressed as Z_k (k = 1, 2, . . ., *m*) in terms of the multivariate data consisting of *M* variables (*M* = number of variables) listed below (Eq. 1.3). The first principal component is defined as the first principal component Z_1, which corresponds to a line that passes through the multidimensional mean and minimizes the sum of squares of the distances of the points from the line in *M*-dimensional space. The second principal component corresponds to the same concept after all correlation with the first principal component has been subtracted from the points. Each succeeding component accounts for as much of the remaining variability as possible. The number of principal components is less than or equal to the number of original variables ($m \leq M$). This transformation is defined such that the first principal component has the largest possible variance (i.e., accounts for as much of the variability in the data as possible). In turn, each succeeding component has the highest variance possible under the constraint that it be orthogonal to (i.e., uncorrelated with) the preceding components.

$$Z_1 = a_{11}X_1 + \ldots + a_{1j}X_j + \ldots + a_{1M}X_M$$

$$\ldots$$

$$Z_j = a_{j1}X_1 + \ldots + a_{jj}X_j + \ldots + a_{jM}X_M \tag{1.3}$$

$$\ldots$$

$$Z_m = a_{m1} X_1 + \dots + a_{mj} X_j + \dots + a_{mM} X_M$$

The results of PCA are usually reported as a complementary set of principal component score, contribution ratio, and component loadings. Component loadings are the correlation coefficients between the variables (rows of j) and components (columns of k), represented as $r (X_j, Z_k)$. The contribution ratio (Eq. 1.4) is obtained by extracting, from a set of original multivariate data, a reduced set of k components that accounts for most of the variance in the original variables. A higher contribution rate reflects larger original data.

$$\%Var[Z_k] = \frac{V[Z_k]}{\sum_{u=1}^{M} V[X_u]} \cdot 100 \tag{1.4}$$

1.4.6 Results

Table 1.2 shows the mean SNP frequencies for a heterozygous genotype. This analysis was performed on both the SNPs obtained from our previous study and those from the HapMap database. The results showed more significant differences in the mean SNP (G/T) frequencies for heterozygous alleles within the SNPs from our previous study.

Tables 1.3 and 1.4 and Figs. 1.6–1.9 present the diagnostic metrics for the more significant components of the four kinds of PCA results (I, II, III, and IV, as described above), including the correlation matrix, principal component loadings, and principal component scores.

The correlation matrix results showed very high correlations between SNP (G/T) with the seasonal UVR at 310 nm. The UVR in autumn and that in winter were more highly correlated to SNP (G/T) than those in spring and summer. PCAs were performed on both the SNPs obtained from our previous study and those from the HapMap database; the correlation results were higher for the SNPs from our previous study. In contrast to SNP (G/T), the major UVR variables were negatively correlated to both SNP (C/T) and SNP (A/G) variables in both cases I and II such that an increase in the major UVR variable(s) corresponded to decreases in both SNP (C/T) and SNP (A/G) variables.

Table 1.2   Mean SNP frequencies for heterozygous alleles in autumn

Environment	UVR (310 nm, Autumn)	123.523	26.708	51.715
	20 SNPs in 7 genes	Africans	Europeans	East Asians
Genome	SNP (C/T)	0.247	0.345	0.323
Genome	SNP (A/G)	0.206	0.208	0.399
Genome	SNP (G/T)	0.405	0.093	0.131
Genome	SNP (A/T)	0.545	0.481	0.413
Genome	SNP (A/C)	0.367	0.100	0.500

	553 SNPs in *OCA2* gene	Africans	Europeans	East Asians
Genome	SNP (A/C)	0.337	0.299	0.180
Genome	SNP (A/G)	0.260	0.234	0.187
Genome	SNP (A/T)	0.199	0.271	0.157
Genome	SNP (C/G)	0.292	0.246	0.153
Genome	SNP (C/T)	0.279	0.241	0.208
Genome	SNP (G/T)	0.251	0.155	0.137

Table 1.3   The correlation matrix in cases I and II

	UVR (Autumn)	UVR (Summer)	UVR (Winter)	UVR (Spring)
UVR (Autumn)	1.00000	0.90878	0.99078	0.99573
UVR (Summer)	0.90878	1.00000	0.84387	0.94341
UVR (Winter)	0.99078	0.84387	1.00000	0.97405
UVR (Spring)	0.99573	0.94341	0.97405	1.00000
SNP (C/T)	−0.99905	−0.88973	−0.99574	−0.99076
SNP (A/G)	−0.27792	0.14828	−0.40549	−0.18807
SNP (G/T)	0.99023	0.84172	0.99999	0.97314
SNP (A/T)	0.70147	0.34009	0.79156	0.63270
SNP (A/C)	0.42726	0.76556	0.30084	0.50888

Principal component loading results showed a high first principal component loading in SNP (G/T), which was positively correlated to the SNP (G/T) variable in case I (Table 1.4). The other principal component loadings were low in other SNPs and not correlated to other SNPs in cases II, III, and IV. This indicates a strong relationship between SNP (G/T) and autumn.

Table 1.4   The first principal component loading in case I

	PC1
UVB (Autumn)	0.94670
SNP (C/T)	−0.95984
SNP (A/G)	−0.57253
SNP (G/T)	0.98236
SNP (A/T)	0.89366
SNP (A/C)	0.11325

The biplots of these PCAs show the principal component scores of the human racial population represented as points on the first two principal component axes (Figs. 1.6–1.9). The points of the principal component scores of the human racial population are distributed separately, dividing the data into two groups, one group of Africans and one group of Europeans and East Asians in cases I, II, and IV.

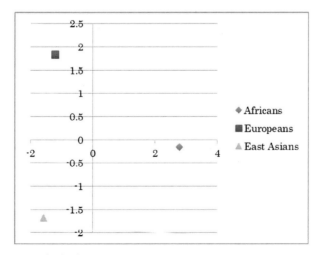

Figure 1.6   The biplot in case I.

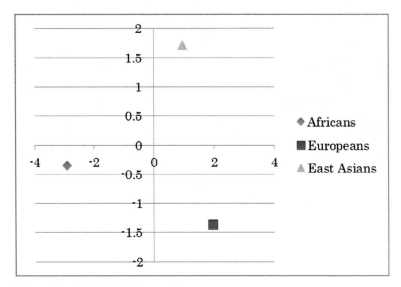

Figure 1.7   The biplot in case II.

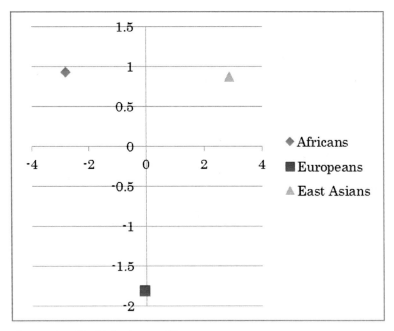

Figure 1.8   The biplot in case III.

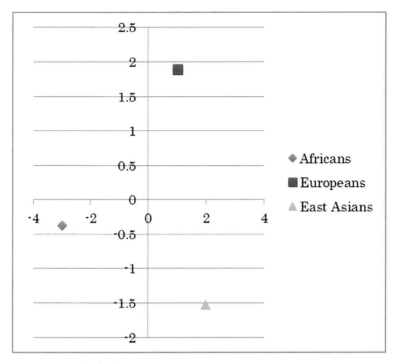

Figure 1.9   The biplot in case IV.

1.4.7   Discussion

A large database or big data set is expected to be required for multidimensional analysis; therefore, our method included PCA. Skin pigmentation is determined by germ-line DNA, and genetic mutations are passed on to the next generation by an unidentified mechanism, leading to mutations of skin pigmentation. Our results suggest the possibility that skin pigmentation might be subject to UVR-induced mutations.

We identified significant differences in the mean SNP (G/T) frequencies for heterozygous alleles and in all seasonal UVR values among the human racial populations. The biplots from the PCAs show that the subjects were divided into two groups: one group of Africans and other group of Europeans and East Asians. The first principal component scores were positive for Africans and negative for Europeans and East Asians in case I, while the first principal component score was negative for Africans and positive for Europeans and East Asians in case II. Furthermore, all PCA results show strong positive correlations between the major seasonal UVR variable(s) and SNP (G/T) compared to other SNPs. Stronger correlations were especially found within the data set obtained with SNPs from our previous study. These findings suggest that SNP (G/T) and the seasonal UVR at 310 nm contribute to the separation of Africans and Europeans/East Asians.

Zhang's research on DNA mutation patterns of human ribosomal protein pseudogene sequences has revealed that nucleotide transitions (C:G→T:A or T:A→C:G) are more common than transversions [61]. Between the two transitional events, C:G→T:A was much more frequent than T:A→C:G [61]. The former case can be caused by a change in the methylation of the cytosine bases and has been known to have a larger substitution rate than in other base substitutions. Between the two transversional events, G→T and T→G, neither occurred more frequently than others and they show no significant characteristics. However, our results indicate very high correlations between SNP (G/T) frequencies for skin pigmentation with the seasonal UVR at 310 nm. It is possible that the whole genome does not have significant characteristics, but the genes for skin pigmentation may have such tendency.

The available data also indicate that mutations may occur due to 8-oxoguanine (8-oxoG). Among the various forms of DNA damage, oxidative DNA lesions caused by reactive oxygen species (ROS)—which are generated both as a by-product of oxidative metabolism and as a consequence of exposure to ionizing radiation and other environmental factors—are considered to be a major threat to the genome [62, 63]. Among the four bases, guanine is known to be the most susceptible to oxidation, and its simple oxidized form, 8-oxoG, is a major oxidation product in DNA or nucleotides [64]. 8-oxoG is a potent premutagenic lesion because it can pair with adenine as well as cytosine during DNA replication and thus can cause a G:C to T:A transversion mutation [65]. It is thus possible that 8-oxoG formation, which is involved in active oxygen through UV exposure, may contribute to G:C to T:A transversion mutations. The mutations may then be passed on to the next generation, leading to skin pigmentation mutations and resulting in skin pigmentation variations, supporting the theory that depigmented and tannable skin may have evolved numerous times in hominin evolution via independent genetic pathways under natural selection.

1.5 Conclusions

Analyzing the relationships between SNP frequencies for heterozygous genes and the seasonal UVR levels as the environment changes will enable elucidation of the nature of the selective mechanisms involved in the UVR-induced evolution of skin pigmentation. Skin pigmentation fulfills the criteria for a successful evolutionary gene–environment interaction model. First, it was produced by an imperfect replicator [48]; skin pigmentation is determined by germ-line DNA, which is subject to mutations induced by UVR. Second, natural selection occurs through differential survival and reproduction rates of different phenotypes under different UVR levels. Furthermore, skin pigmentation is an attractive model system for understanding and teaching evolutionary gene–environment interactions because it is readily visible and the basic contributory mechanisms are easily understood.

Europeans who are born and reach adulthood in Europe and then migrate from Europe to Australia appear to be at one-fourth of the risk of developing melanoma compared to Europeans who were born and brought up in Australia. On the other hand, the Europeans who were born in Europe and migrated from Europe to Australia during childhood exhibited similar rates of melanoma development as the Europeans who were born and brought up in Australia. This may indicate that intense UV exposure in childhood and adolescence is a causative factor in melanoma development [66]. It further suggests that Europeans who are born and brought up in regions with high altitudes have a low ability to adapt to the strong UVR at low latitudes. Therefore, it is important to further investigate human adaptability to UVR in order to enable better prediction of health risks caused by harsh environmental conditions and the construction of preventive interventions.

Despite the importance of gene–environment interactions for complex phenotypes, there has been little progress in developing methods for detecting the interactions involved in skin color variation among human racial populations. In this chapter, we have proposed a spatial statistical approach that provides a useful and powerful theoretical framework for investigating the adaptation mechanisms involved in skin color variation by detecting gene–environment interactions. The complexity of spatial data and intrinsic spatial relationships limits the usefulness of conventional data-mining techniques for extracting spatial patterns. Spatial statistical analysis enables the discovery of interesting, previously unknown, potentially useful patterns from spatial databases. Efficient tools for extracting information from geospatial data are crucial to organizations that make decisions based on spatial data sets, and are necessary in a variety of fields, including ecology and environmental management, public safety, transportation, Earth science, epidemiology, and climatology [67, 68].

Several factors must be considered in trying to clarify the evolutionary adaptations of skin color variation that resulted from localized adaptation to various UVR conditions through natural selection. First, information on genetic and environmental factors will have to be integrated into one database. Second, the evolution of the genome occurs through the accumulation of environmental association rules that correspond to the cumulative knowledge the genome has acquired regarding its environment [53]. Third, when UVR is the environmental factor, GIS can serve as a medium for analyzing and visualizing spatial patterns. Finally, the identification of functional relationships between molecular evolutionary genetics (SNP frequencies for a heterozygous gene) and seasonal UVR measurements requires computational methods that address spatial data sets.

Our PCA-based approach to spatial statistical analysis is a convenient means of applying association rules to extract spatial associations between genetic and environmental variables that determine skin color variation for different human populations. This approach may reveal environment-based differences in genetic traits and can maximize the use of available data to improve the understanding of the adaptive mechanisms that give rise to human skin color variations. Spatial statistical analysis may identify biological mechanisms that regulate human adaptability to UVR. Moreover, identifying the complex mechanisms that have shaped evolutionary changes in skin color may advance our understanding of the history of human adaptation to local environments, potentially having important public health implications. Understanding human adaptability to UVR may enable prediction of the impact of environmental changes on health risks, particularly regarding high UV exposure secondary to ozone depletion.

Traditional parametric statistical approaches have limited power for modeling the high-order, nonlinear interactions that are probably important in generating complex phenotypes. Given the newly developed remote-sensing technologies that capture more detailed, higher-resolution UV wavelength measurements, as well as emerging approaches to gene–environmental interactions that integrate a coherent biological framework, we believe that the mechanism of human adaptability to UV environments will be clarified over the next several years by studies of skin pigmentation at all organizational levels, from genes to populations, providing a systematic understanding of this phenomenon. New powerful spatial statistical tools and automated learning methods will be developed to computationally model the relationships among genetic variables, protein variables, cell variables, tissue variables, organ variables, physical variables, environmental exposure, and skin color variation.

References

1. Anno, S., Abe, T., Sairyo, K., Kudo, S., Yamamoto, T., Ogata, K., and Goel, V. K. (2007). Interactions between SNP alleles at multiple loci and variation in skin pigmentation in 122 Caucasians. *Evol. Bioinform. Online*, **3**, pp. 169–178.

2. Aoki, K. (2002). Sexual selection as a cause of human skin color variation: Darwin's hypothesis revisited. *Ann. Hum. Biol.*, **29**(6), pp. 589–608.

3. John, P. R., Makova, K., Li, W. H., Jenkins, T., and Ramsay, M. (2003). DNA polymorphism and selection at the melanocortin-1 receptor gene in normally pigmented southern African individuals. *Ann. N. Y. Acad. Sci.*, **994**, pp. 299–306.

4. Akey, J. M., Wang, H., Xiong, M., Wu, H., Liu, W., Shriver, M. D., and Jin, L. (2001). Interaction between the melanocortin-1 receptor and P genes contributes to inter-individual variation in skin pigmentation phenotypes in a Tibetan population. *Hum. Genet.*, **108**(6), pp. 516–520.

5. Rees, J. L. (2003). Genetics of hair and skin color. *Annu. Rev. Genet.,* **37**, pp. 67–90.

6. Flanagan, N., Healy, E., Ray, A., Philips, S., Todd, C., Jackson, I. J., Birch-Machin, M. A., and Rees, J. L. (2000). Pleiotropic effects of the melanocortin 1 receptor (*MC1R*) gene on human pigmentation. *Hum. Mol. Genet.,* **9**(17), pp. 2531–2537.

7. Peng, S., Lu, X. M., Luo, H. R., Xiang-Yu, J. G., and Zhang, Y. P. (2001). Melanocortin-1 receptor gene variants in four Chinese ethnic populations. *Cell Res.,* **11**(1), pp. 81–84.

8. Voisey, J., Box, N. F., and Van Daal, A. (2001). A polymorphism study of the human agouti gene and its association with *MC1R*. *Pigment Cell Res.,* **14**(4), pp. 264–267.

9. Nakayama, K., Fukamachi, S., Kimura, H., Koda, Y., Soemantri, A., and Ishida, T. (2002). Distinctive distribution of AIM1 polymorphism among major human populations with different skin color. *J. Hum. Genet.,* **47**(2), pp. 92–94.

10. Naysmith, L., Waterston, K., Ha, T., Flanagan, N., Bisset, Y., Ray, A., Wakamatsu, K., Ito, S., and Rees, J. L. (2004). Quantitative measures of the effect of the melanocortin 1 receptor on human pigmentary status. *J. Invest. Dermatol.,* **122**(2), pp. 423–428.

11. Myles, S., Somel, M., Tang, K., Kelso, J., and Stoneking, M. (2007). Identifying genes underlying skin pigmentation differences among human populations. *Hum Genet.,* **120**, pp. 613–621.

12. Izagirre, N., Garcia, I., Junquera, C., de la Rúa, C., and Alonso, S. (2006). A scan for signatures of positive selection in candidate loci for skin pigmentation in humans. *Mol. Biol. Evol.,* **23**(9), pp. 1697–1706.

13. McEvoy, B., Beleza, S., and Shriver, M. D. (2006). The genetic architecture of normal variation in human pigmentation: an evolutionary perspective and model. *Hum. Mol. Genet.,* **15**(2), pp. R176–R181.

14. Bonilla, C., Boxill, L. A., Donald, S. A., Williams, T., Sylvester, N., Parra, E. J., Dios, S., Norton, H. L., Shriver, M. D., and Kittles, R. A. (2005). The 8818G allele of the agouti signaling protein (ASIP) gene is ancestral and is associated with darker skin color in African Americans. *Hum. Genet.,* **116**, pp. 402–406.

15. Makova, K., and Norton, H. (2005). Worldwide polymorphism at the *MC1R* locus and normal pigmentation variation in humans. *Peptides,* **26**, pp. 1901–1908.

16. Naysmith, L., Waterston, K., Ha, T., Flanagan, N., Bisset, Y., Ray, A., Wakamatsu, K., Ito, S., and Rees, J. L. (2004). Quantitative measures of

the effect of the melanocortin 1 receptor on human pigmentary status. *J. Invest. Dermatol.,* **122**(2), pp. 423–428.

17. Ancans, J., Flanagan, N., Hoogduijn, M. J., and Thody, A. J. (2003). P-locus is a target for the melanogenic effects of MC-1R signaling: a possible control point for facultative pigmentation. *Ann. N. Y. Acad. Sci.,* **994**, pp. 373–377.

18. Rees, J. L. (2003). Genetics of hair and skin color. *Annu. Rev. Genet.,* **37**, pp. 67–90.

19. Tadokoro, T., Yamaguchi, Y., Batzer, J., Coelho, S. G., Zmudzka, B. Z., Miller, S. A., Wolber, R., Beer, J. Z., and Hearing, V. J. (2005). Mechanisms of skin tanning in different racial/ethnic groups in response to ultraviolet radiation. *J. Invest. Dermatol.,* **124**, pp. 1326–1332.

20. Bonilla, C., Shriver, M. D., Parra, E. J., Jones, A., and Fernández, J. R. (2004). Ancestral proportions and their association with skin pigmentation and bone mineral density in Puerto Rican women from New York City. *Hum. Genet.,* **115**, pp. 57–68.

21. Shriver, M. D., Parra, E. J., Dios, S., Bonilla, C., Norton, H., Jovel, C., Pfaff, C., Jones, C., Massac, A., Cameron, N., Baron, A., Jackson, T., Argyropoulos, G., Jin, L., Hoggart, C. J., McKeigue, P. M., and Kittles, R. A. (2003). Skin pigmentation, biogeographical ancestry, and admixture mapping. *Hum. Genet.,* **112**, pp. 387–399.

22. Hoggart, C. J., Parra, E. J., Shriver, M. D., Bonilla, C., Kittles, R. A., Clayton, D. G., and McKeigue, P. M. (2003). Control of confounding of genetic associations in stratified populations. *Am. J. Hum. Genet.,* **72**, pp. 1492–1504.

23. Sturm, R. A., Teasdale, R. D., and Box, N. F. (2001). Human pigmentation genes: identification, structure and consequences of polymorphic variation. *Gene,* **277**, pp. 49–62.

24. National Library of Medicine. *Searchable NCBI site for Single Nucleotide Polymorphisms,* http://www.ncbi.nlm.nih.gov/projects/SNP/.

25. Kokoris, M., Dix, K., Moynihan, K., Mathis, J., Erwin, B., Grass, P., Hines, B., and Duesterhoeft, A. (2000). High-throughput SNP genotyping with the Masscode system. *Mol. Diagn.,* **5**(4), pp. 329–340.

26. Ogata, K., Ikeda, S., and Ando, E. (2002). QIAGEN Genomics Inc. A study of SNP genotyping using Masscode™ technology. *Shimadzu Hyoka.,* **58**, pp. 125–129.

27. Jorde, L. B. (2000). Linkage disequilibrium and the search for complex disease genes. *Genome Res.,* **10**, pp. 1435–1444.

28. Pritchard, J. K., and Przeworski, M. (2001). Linkage disequilibrium in humans: models and data. *Am. J. Hum. Genet.,* **69**, pp. 1–14.

29. Zhao, J., Jin, L., and Xiong, M. (2006). Test for interaction between two unlinked loci. *Am. J. Hum. Genet.,* **79**(5), pp. 831–845.

30. Lewontin, R. C. (1964). The interaction of selection and linkage. I. General considerations; heterotic models. *Genetics,* **49**, pp. 49–67.

31. Terwilliger, J., and Ott, J. (1994). *Handbook of Human Genetic Linkage* (Johns Hopkins University Press, Baltimore).

32. Barrett, J. C., Fry, B., Maller, J., and Daly, M. J. (2005). Haploview: analysis and visualization of LD and haplotype maps. *Bioinformatics,* **21**, pp. 263–265.

33. Warnes, G., and Leisch, F. (2005). Genetics: population genetics [Computer program]. R package version 1.2.0, http://cran.r-project. org/src/contrib/PACKAGES.html.

34. Anno, S., Abe, T., and Yamamoto, T. (2008). Interactions between SNP alleles at multiple loci contribute to skin color differences between Caucasoid and Mongoloid subjects. *Int. J. Biol. Sci.,* **4**, pp. 81–86.

35. Libby, R. T., Lillo, C., Kitamoto, J., Williams, D. S., and Steel, K. P. (2004). Myosin Va is required for normal photoreceptor synaptic activity. *J. Cell Sci.,* **117**, pp. 4509–4515.

36. Sabeti, P. C., Reich, D. E., Higgins, J. M., Levine, H. Z., Richter, D. J., Schaffner, S. F., Gabriel, S. B., Platko, J. V., Patterson, N. J., McDonald, G. J., Ackerman, H. C., Campbell, S. J., Altshuler, D., Cooper, R., Kwiatkowski, D., Ward, R., and Lander, E. S. (2002). Detecting recent positive selection in the human genome from haplotype structure. *Nature,* **419**, pp. 832–837.

37. Alonso, S., Izagirre, N., Smith-Zubiaga, I., Gardeazabal, J., Díaz-Ramón, J. L., Díaz-Pérez, J. L., Zelenika, D., Boyano, M. D., Smit, N., and de la Rúa, C. (2008). Complex signatures of selection for the melanogenic loci TYR, TYRP1 and DCT in humans. *BMC Evol. Biol.,* **8**(74), pp. 1–14.

38. Lao, O., de Gruijter, J. M., van Duijn, K., Navarro, A., and Kayser, M. (2007). Signatures of positive selection in genes associated with human skin pigmentation as revealed from analyses of single nucleotide polymorphisms. *Ann. Hum. Genet.,* **71**, pp. 354–369.

39. Librado, P., and Rozas, J. (2009). DnaSP v5: Software for comprehensive analysis of DNA polymorphism data. *Bioinformatics,* **25**, pp. 1451–1452.

40. Edwards, M., Bigham, A., Tan, J., Li, S., Gozdzik, A., Ross, K., Jin, L., and Parra, E. J. (2010). Association of the OCA2 polymorphism His615Arg

with melanin content in East Asian populations: further evidence of convergent evolution of skin pigmentation. *PLOS Genet.,* **6**(3), e1000867.

41. Voight, B. F., Kudaravalli, S., Wen, X., and Pritchard, J. K. (2006). A map of recent positive selection in the human genome. *PLOS Biol.,* **4**, e72.

42. Williamson, S. H., Hubisz, M. J., Clark, A. G., Payseur, B. A., Bustamante, C. D., and Nielsen, R. (2007). Localizing recent adaptive evolution in the human genome. *PLOS Genet.,* **3**, e90.

43. Sabeti, P. C., Varilly, P., Fry, B., Lohmueller, J., Hostetter, E., Cotsapas, C., Xie, X., Byrne, E. H., McCarroll, S. A., Gaudet, R., Schaffner, S. F., Lander, E. S., and International HapMap Consortium. (2007). Genome-wide detection and characterization of positive selection in human populations. *Nature,* **449**, pp. 913–918.

44. Johansson, A., and Gyllensten, U. (2008). Identification of local selective sweeps in human populations since the exodus from Africa. *Hereditas,* **145**, pp. 126–137.

45. Norton, H. L., Kittles, R. A., Parra, E., McKeigue, P., Mao, X., Cheng, K., Canfield, V. A., Bradley, D. G., McEvoy, B., and Shriver, M. D. (2007). Genetic evidence for the convergent evolution of light skin in Europeans and East Asians. *Mol. Biol. Evol.,* **24**(3), pp. 710–722.

46. Jablonski, N. G., and Chaplin, G. (2000). The evolution of human skin coloration. *J. Hum. Evol.,* **39**, pp. 57–106.

47. Chaplin, G. (2004). Geographic distribution of environmental factors influencing human skin coloration. *Am. J. Phys. Anthropol.,* **125**, pp. 292–302.

48. Jablonski, N. G., and Chaplin, G. (2010). Human skin pigmentation as an adaptation to UV radiation. *Proc. Natl. Acad. Sci. U. S. A.,* **107**, pp. 8962–8968.

49. Loomis, W. F. (1967). Skin-pigment regulation of vitamin-D biosynthesis in man. *Science,* **157**, pp. 501–506.

50. Murray, F. G. (1934). Pigmentation, sunlight, and nutritional disease. *Am. Anthropol.,* **36**, pp. 438–445.

51. Dunkern, T. R., Fritz, G., and Kaina, B. (2001). Ultraviolet light-induced DNA damage triggers apoptosis in nucleotide excision repair-deficient cells via Bcl-2 decline and caspase-3/-8 activation. *Oncogene,* **20**, pp. 6026–6038.

52. Anno, S., Ohshima, K., Abe, T., Tadono, T., Yamamoto, A., and Igarashi, T. (2013). Approaches to detecting gene-environment interactions in human variation using genetic engineering, remote sensing and GIS. *Journal of Earth Science and Engineering,* **3**(6), pp. 371–378.

53. Irizarry, K. J. L., Merriman, B., Bahamonde, M. E., Wong, M-L., and Licinio, J. (2005). The evolution of signaling complexity suggests a mechanism for reducing the genomic search space in human association studies. *Mol. Psychiatry,* **10**, pp. 14–26.

54. Murcray, C. E., Lewinger, J. P., and Gauderman, W. J. (2009). Gene–environment interaction in genome-wide association studies. *Am. J. Epidemiol.,* **169**(2), pp. 219–226.

55. Duell, E. J., Bracci, P. M., Moore, J. H., Burk, R. D., Kelsey, K. T., and Holly, E. A. (2008). Detecting pathway-based gene–gene and gene–environment interactions in pancreatic cancer. *Cancer Epidemiol. Biomarkers Prev.,* **17**(6), pp. 1470–1479.

56. McEachin, R. C., Keller, B. J., Saunders, E. F., and McInnis, M. G. (2008). Modeling gene-by-environment interaction in comorbid depression with alcohol use disorders via an integrated bioinformatics approach. *BioData Min.,* **1**, pp. 1–13.

57. Chien, Y. (1978). Interactive pattern recognition. In *Electrical Engineering and Electronics III*, pp. 25–64, Thurston, M. O., and Middendorf, W. (eds.) (Marcel Dekker, New York).

58. Jolliffe, I. T. (1986). *Principal Component Analysis* (Springer-Verlag, New York).

59. Kanaya, S., Kudo, Y., Nakamura, Y., and Ikemura, T. (1996). Detection of genes in Escherichia coli sequences determined by genome projects and prediction of protein production levels, based on multivariate diversity in codon usage. *CABIOS,* **12**(3), pp. 213–225.

60. Quackenbush, J. (2001). Computational analysis of microarray data. *Nat. Rev. Genet.,* **2**(6), pp. 418–427.

61. Zhang, Z., and Gerstein, M. (2003). Patterns of nucleotide substitution, insertion and deletion in the human genome inferred from pseudogenes. *Nucleic Acids Res.,* **31**(18), pp. 5338–5348.

62. Hanawalt, P. C. (1998). Genomic instability: environmental invasion and the enemies within. *Mutat. Res.,* **400**, pp. 117–125.

63. Barnes, D. E., and Lindahl, T. (2004). Repair and genetic consequences of endogenous DNA base damage in mammalian cells. *Annu. Rev. Genet.,* **38**, pp. 445–476.

64. Kasai, H., and Nishimura, S. (1984). Hydroxylation of deoxyguanosine at the C-8 position by ascorbic acid and other reducing agents. *Nucleic Acids Res.,* **12**, pp. 2137–2145.

65. Shibutani, S., Takeshita, M., and Grollman, A. P. (1991). Insertion of specific bases during DNA synthesis past the oxidation-damaged base 8-oxodG. *Nature,* **349**, pp. 431–434.

66. Armstrong, B. K., and Kricker, A. (1993). Sun exposure causes both nonmelanocytic skin cancer and malignant melanoma. In *Proceedings on Environmental UV Radiation and Health Effects*, pp. 105–113, Schpka, H-J., and Steinmetz, M. (eds.).

67. Roddick, J. F., and Spiliopoulou, M. (1999). A bibliography of temporal, spatial and spatio-temporal data mining research. *SIGKDD Explorations,* **1**(1), pp. 34–38.

68. Shekhar, S., and Chawla, S. (2003). *Spatial Databases: A Tour* (Prentice Hall, New Jersey).

Chapter 2

Information Theoretic Methods for Gene–Environment Interaction Analysis

Jonathan Knights and Murali Ramanathan

*Department of Pharmaceutical Sciences, State University of New York,
Buffalo, NY 14214-8033, USA*

jonathanknights.3783@gmail.com, murali@buffalo.edu

2.1 Introduction

From the accumulating evidence provided by sequencing and genome-wide association studies of cancer, cardiovascular disease, and autoimmune disease, it is clear that the pathogenesis of complex diseases and the response to drug therapy are the result of interactions among many exogenous and endogenous factors operating on one or more biological pathways. However, reliably identifying the key underlying processes and translating the findings to clinically actionable decisions have proven difficult [1, 2]. Analyzing, interpreting, and gaining an in-depth understanding of personal genomes and epidemiological study data containing numerous genotypes and multiple environmental exposures present

Gene–Environment Interaction Analysis: Methods in Bioinformatics and Computational Biology
Edited by Sumiko Anno

many critical challenges. However, it is important to overcome these challenges to enable better prevention and therapeutic strategies.

Although this chapter focuses on gene–gene interaction (GGI) and gene–environment interaction (GEI) analysis, the concept of an interaction is ubiquitous and important in many scientific disciplines ranging from economics, sociology, and physics to the biomedical and pharmaceutical sciences. Each field, however, uses different terminology and methodologies for detecting and quantitating interactions. Epistasis in genetics and drug synergy in the pharmaceutical sciences, are examples of interactions.

Broadly defined, an interaction is an outcome that occurs when two or more predictor variables are present together. Biological interactions are frequently mediated by noncovalent binding associations between (or among) various biomolecular species ranging from ions, small molecules, and proteins to RNA and DNA. However, biological interactions can also involve covalent binding associations and may be triggered by, or require, an exchange of energy (e.g., light, radiation, heat, membrane potential, etc.) or information (e.g., electrical signals from nerves or stress from interpersonal interactions).

In contrast, statistical interactions are inferred from available data. Importantly, the genetic variation data used for GGI/GEI analyses are generally biologically distal from proteins and other biomolecules involved in the biological interactions. The statistical evidence for interactions is obtained by assessing the associations of dependent variables with the presence of two or more predictor variables. From this perspective, not all biological interactions will have detectable signatures on data. Operationally, in linear statistical regression analysis, interactions are incorporated as product terms in the regression equation.

Because simultaneous characterization of numerous genetic variations and environmental variables is possible with technologies such as next-generation sequencing, microarrays, and mass spectrometry, data sets for GGI and GEI analysis can potentially involve as many 10^4–10^7 variables. For such a large number of variables, finding potential interactions and defining the best set of those interactions may require distinct methodologies. Therefore, GEI analysis in large data sets should have three major components: (i) metric definition, (ii) interaction detection, and (iii) model synthesis. Methods based on information theory offer a powerful and versatile framework for each of these aspects. This chapter will present an overview of information theoretic metrics and their biomedical research applications in GEI analysis.

2.2 Information Theoretic Metrics and Searching for GEIs

Information theoretic methods are among the most promising approaches for GEI analysis [3–6] because the underlying metrics have a well-developed theory and are versatile.

2.2.1 Entropy and Mutual Information

2.2.1.1 Entropy

The fundamental building block of all information theory methods is entropy. Entropy was originally developed by Claude Shannon in his work *The Mathematical Theory of Communication* [7] for quantitating the amount of information carried in a message. However, entropy metrics are applicable to a much wider range of problems involving random variables.

The Shannon entropy $H(X)$ of a random variable X is defined as

$$H(X) = -\sum_{x \in X} p(x) \log p(x) \qquad (2.1)$$

where $p(x)$ is the probability of each of the possible states, x, of the random variable X. Shannon's original definition used bits, which are obtained by using logarithms to the base 2 ($\log_2$) as the measure of information in Eq. 2.1. Using a logarithm to a different base simply provides an entropy measure that differs by a constant of proportionality.

The entropy defined by Eq. 2.1 is always positive: $H(X) \geq 0$. The entropy is zero when there is no uncertainty regarding the outcome, for example, in a Bernoulli distribution with $p(x) = 1$. The entropy is maximal when all of the states are equiprobable. Shannon entropy can also be viewed as the expectation of $-\log p(x)$. In lay terms, Shannon entropy can be viewed as a measure of the information gained when an uncertain event occurs.

Although Shannon entropy shares striking similarities in many mathematical properties with thermodynamic entropy in physics, engineering, and chemistry, there are also some nuanced differences. Thermodynamic entropy is viewed as the amount of disorder in a macroscopic system and the probabilities for entropy calculations are based on the equilibrium distribution between the possible states. Notably, thermodynamic entropy is an extensive measure whose value is proportional to the size of the system.

In epidemiological and pharmacogenomics studies, the relationship of the genetic and environmental predictors, X, to the response variable or phenotype, Y, is of primary interest. Shannon entropy can be extended to two random variables, X and Y, as the joint entropy, $H(X,Y)$, which is defined as

$$H(X,Y) = -\sum_{x,y} p(x,y)\log p(x,y) \qquad (2.2)$$

The summation is now taken over all of the states (x,y) in the joint distribution of (X,Y).

2.2.1.2 Mutual information

Mutual information, $I(X,Y)$, is an information theoretic metric that measures the amount of information one variable carries about another. Although the concept of mutual information is embedded in Shannon's work [7], McGill [8] described its properties and interpretation in detail. Mutual information is defined as

$$I(X,Y) = H(X,Y) - H(X) - H(Y) \qquad (2.3)$$

If two variables are independent, then knowing the value of one variable gives us no information about the second: the mutual information $I(X,Y)$ is zero when X and Y are independent. Mutual information is maximal when X and Y are completely dependent, that is, one variable completely describes the other. The mutual information is proportional to the log-likelihood relative to independence. The larger the mutual information, the more informative the variable is about the phenotype. In statistical terminology, mutual information can be viewed as assessing the main effects, which according to our information theoretic approach is a first-order interaction.

The Kullback–Leibler divergence (KLD) between two probability mass functions $p(x)$ and $q(x)$ is denoted by KLD$(p||q)$ and is also known as the relative entropy. The definition of the KLD is [9]

$$KLD(p\,||\,q) = \sum_{x \in X} p(x) \log \frac{p(x)}{q(x)} \tag{2.4}$$

The KLD measures the inefficiency of assuming that the distribution is q when the true distribution is p. The KLD always takes nonnegative values and is zero only if $p = q$ [10]. If the distribution p can be viewed as representing a statistical hypothesis, the KLD is the expected log-likelihood ratio.

We now discuss how entropy-based metrics can be used to search, identify, and measure GGIs and GEIs.

2.2.1.3   The *k*-way interaction information

The analysis of GGI and GEI requires multivariate extensions of entropy-based metrics. The *k*-way interaction information (KWII) is a multivariate information theoretic measure that quantitates the information that can only be obtained about the phenotype of interest from specific subsets of variables in a given data set [8, 11].

For the three-variable case (A,B,Y), where A and B are genetic or environmental predictor variables and Y is the phenotype of interest, the KWII is written in terms of the individual entropies $H(A)$, $H(B)$, and $H(Y)$ and of the joint entropies $H(A,B)$, $H(A,Y)$, $H(B,Y)$, and $H(A,B,Y)$ as

$$\begin{aligned} \mathrm{KWII}(A,B,Y) = {}&- H(A) - H(B) - H(Y) + H(A,B) + H(A,Y) \\ &+ H(B,Y) - H(A,B,Y) \end{aligned} \tag{2.5}$$

Figure 2.1 is an information Venn diagram that highlights the relationship of the KWII in the three-variable case to the entropies of the lower-order subsets. The central core of the Venn diagram represents the KWII.

For the general case on the set $v = \{X_1, X_2, \ldots, X_k, Y\}$, containing k predictors and the phenotype, Y, the KWII is written as an alternating sum over all possible subsets T of v. Using the difference-operator notation of Han [12]

$$\mathrm{KWII}(v) = -\sum_{T \subseteq v} (-1)^{|v|-|T|} H(T) \tag{2.6}$$

The symbols, $|v|$ and $|T|$, represent the size of the set v and its subset T, respectively. The number of genetic and environmental variables, k (not including the phenotype), in a combination is called the *order* of the interaction.

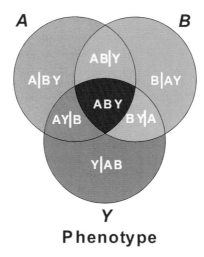

Phenotype

Figure 2.1 A Venn diagram representing the KWII terms for the three-variable case. The predictor variables are denoted by *A* and *B*; the phenotype is denoted by *Y*. The information in the central region marked *ABY* corresponds to the magnitude of the KWII. Adapted from Ref. [16].

The KWII quantifies interactions by representing the information that cannot be obtained without observing all k predictor variables and the phenotype, *Y*, at the same time [8, 13–15]. The KWII of a given combination of variables is a parsimonious interaction metric in that it does not contain contributions arising from the KWII of lower order combinations (i.e., the subsets of the k-way variable combination).

For a single variable, the KWII is the entropy; for the two-variable situation (e.g., a single genetic or environmental predictor and the phenotype), the KWII is the mutual information. Therefore, in the bivariate case, KWII ≥ 0. However, when there are more than two variables, the KWII can be positive, negative, or zero.

We operationally define *interaction* as follows: "For each variable combination containing the phenotype, a positive KWII value indicates the presence of an interaction, negative values of KWII indicate the presence of redundancy, and a KWII value of zero suggests the lack of informative k-way interactions."

The KWII has been previously demonstrated as an effective metric for quantitating GGI and GEI [11, 17–19].

2.2.1.4 Total correlation information

The total correlation information (TCI) is a multivariate information theoretic metric that provides the foundation for the phenotype-associated information, which is useful for developing GEI and GGI search algorithms.

For the same three-variable case in Eq. 2.5, the TCI [20] is defined as

$$TCI(A,B,Y) = H(A) + H(B) + H(Y) - H(A,B,Y) \qquad (2.7)$$

For the general k-variable case on the set $\upsilon = \{X_1,X_2,...,X_k,Y\}$, containing k predictors and the phenotype, Y, the TCI is generalized to

$$TCI(\upsilon) = H(Y) + \sum_{i=1}^{k} H(X_i) - H(\upsilon) \qquad (2.8)$$

The TCI represents the difference between the sum of the individual entropies (represented by the $H(Y)$ and $H(X_i)$ terms) and the joint entropy, $H(\upsilon)$. The TCI is a general measure of correlation between a set of variables: If the variables in the set are independent, TCI = 0.

2.2.1.5 Phenotype-associated information

The phenotype-associated information (PAI) is derived from the TCI and quantitates the amount of information a set of variables carries about the phenotype alone [18]. This is accomplished by removing the TCI of the set υ, defined above, containing the predictors and the phenotype, from the set $\chi = \{X_1,X_2,...,X_k\}$, containing the predictors alone. For the k-variable case

$$PAI(\upsilon) = TCI(\upsilon) - TCI(\chi) \qquad (2.9)$$

Since the TCI is a measure of interdependency among the variables, the PAI measures the dependency of the phenotype on the set of predictors after removing the dependencies among the set of predictors. Dependencies among genetic and environmental variables can arise from multiple sources, such as linkage disequilibrium (LD), population admixture/heterogeneity, or exposure to multiple environmental pollutants. The ability of the PAI to quantitate only the information carried about the phenotype in a set of predictors has made it a powerful search metric for GEI detection [17, 18].

The PAI is directly related to the KWII of the genetic and environmental variables, and the phenotype, such that

$$\text{PAI}(\upsilon) = \sum_{T \subseteq \chi} \text{KWII}(T, Y) \qquad (2.10)$$

The PAI is the cumulative KWII present in all subset combinations of the variables in χ and Y. The PAI is always positive and increases monotonically with the combination size. When a noninformative predictor is added to a combination, the PAI is unchanged. However, the addition of an informative predictor increases the PAI. These interesting properties of the PAI and its relationship to the KWII enable search algorithms that identify interesting regions of a combinatorial space. The individual KWIIs can then be computed for the reduced combinatorial space. The computational challenges related to combinatorial explosion are reduced because GEI analysis is not conducted exhaustively for every possible combination.

2.3 How and Why Do the KWII and the PAI Measure Statistical Interaction?

In the analysis of data for discrete and categorical variables, the presence of a statistical interaction indicates that the effect of the combination of variables is nonadditive. This implies that the probability of obtaining the phenotype for a variable combination is greater (or less) than expected from the probability of the phenotype for each variable individually. In its mathematical form, epistasis, as defined by Fisher [21], is similar to this definition of statistical interaction [22]. In logistic regression analysis, statistical interactions are assumed to be present if the coefficients for the product terms are significant.

The KWII, TCI, and PAI definitions also assess interactions between variables but from an information theoretic framework, wherein entropies, rather than probabilities, are used. As a starting point for exploring their relationships to statistical interactions, first consider the TCI definition in Eq. 2.7 for three variables. The TCI directly compares the information gained from observing the joint distribution, $H(A,B,C)$, with $H(A) + H(B) + H(C)$—the information gained from observing the variables individually—via a simple subtraction. Thus, the definition of the TCI highlights the relationship between interactions and deviations from additivity.

However, the TCI definition contains contributions from the interdependencies among the variables (see Fig. 2.1): these consist of the information gained when, for example, A and B are observed together without C; B and C are observed together with out A; and A and C are observed without B. The KWII removes these contributions to provide a metric that does not contain information from any of the lower-order interactions. The PAI subtracts the interdependencies among the variables that can be identified in the absence of the phenotype. Such interdependencies arise from confounding factors such as LD and correlated environmental exposures.

The advantage of using information theory is that the mathematical framework for identifying interactions can be reduced to additions and subtractions of entropies. In probabilistic terms, the expressions become more involved. For a detailed analysis of the theoretical properties of information theoretic metrics and their relationship to familiar statistical methods, see Ref. [23].

2.4 Performance of KWII and PAI Search Algorithms on Simulated Data

In the following numerical experiments, we simulated data with known patterns of GEI and examined their relationships with the KWII, TCI, and PAI spectra.

Case Study 1: The underlying GEI model for Case Study 1 is summarized in Fig. 2.2A. The simulated data for Case Study 1 consisted of four environmental variables, $E1$ through $E4$, and six single-nucleotide polymorphism (SNP) variables, SNP 1 through SNP 6. The environmental variables $E1$ and $E2$ were assumed to be associated with the disease phenotype, whereas $E3$ and $E4$ were uninformative. The environmental variables $E1$ and $E3$ were assumed to have two states, low exposure (assigned the value L) and high exposure (assigned the value H), and were treated as categorical. The environmental variables $E2$ and $E4$ were assumed

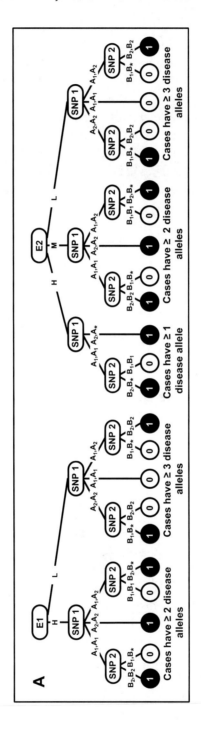

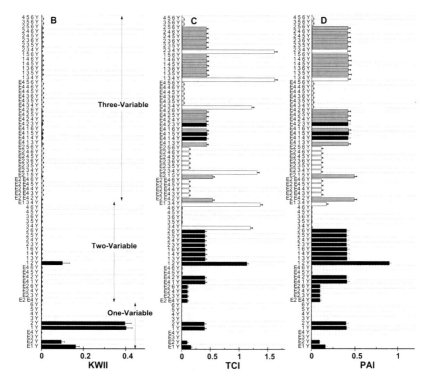

Figure 2.2 (A) The interaction model used to generate the data for Case Study 1. The environmental variables $E1$ (with states H and L) and $E2$ (with states H, M, and L) independently interact with two SNP variables, SNP 1 (with alleles A_1 and A_2) and SNP 2 (with alleles B_1 and B_2), to determine the disease status (controls are indicated by 0, and cases are indicated by 1). The asterisk in a genotype represents a wild card, indicating that either allele is allowable. The uninformative variables are not shown. (B, C, and D) KWII, TCI, and PAI spectra for Case Study 1, respectively. All the one-variable-containing combinations along with the top 20 two-variable and top 20 three-variable combinations with the highest KWII values are shown. The environmental variables are shown as $E1$–$E4$, the SNP variables are numbered 1 to 6, and the phenotype is indicated as C. The combinations are indicated on the y axis. The error bars represent the standard deviations. Reprinted with permission from Ref. [18], Copyright 2008 by the Genetics Society of America.

to have three states, low exposure (assigned the value L), medium exposure (assigned the value M), and high exposure (assigned the value H), and were also treated as categorical. The percentage of subjects in low- and high-exposure groups of $E1$ and $E3$ was 50%

each; the percentage of subjects in low-, intermediate-, and high-exposure groups of $E2$ and $E4$ was 33.33% each, respectively. The disease was modeled to occur for various combinations of exposure to the environmental variables $E1$ and $E2$ via interactions with alleles for two SNPs, SNP 1 and SNP 2. To mimic an additive genetic model, the values 1, 2, and 3 were used to represent the homozygous state for the major allele, the heterozygous genotype, and the homozygous state for the minor allele, respectively, for all six SNP variables. The more common and less common (disease) alleles of SNP 1 and SNP 2 were assigned allele frequencies of 0.9 and 0.1, respectively. The other SNP variables, SNP 3 through SNP 6, were uninformative and had allele frequencies of 0.5. All SNPs were assumed to be diallelic, with the three possible genotypes in Hardy–Weinberg equilibrium. A binary phenotype variable, C, representing case (assigned value = 1) or control (assigned value = 0) was used.

The $E1$ and $E2$ variables were assumed to act independently of each other, and the case phenotype value was assigned when combinations of the SNP genotypes and either environmental variable resulted in a case. The SNP variables SNP 3 and SNP 4 were assumed to be in LD ($R^2 = 0.9$) with each other.

The model represents a challenging scenario with environmental heterogeneity, that is, the two different environmental variables increase disease risk independently via the same genetic variables.

2.4.1 KWII as an Interaction Metric

The KWII (Fig. 2.2B), TCI (Fig. 2.2C), and PAI (Fig. 2.2D) spectra, obtained using the PAI-based AMBIENCE search algorithm, are compared to the corresponding results from an exhaustive search (EXS) of all combinations containing four variables or less for Case Study 1. The goal is to assess the effectiveness of the KWII metric by verifying that the critical interactions are identified.

For each method of search, the 20 combinations with the highest KWII values are presented for the one-variable, two-variable, and three-variable combinations. The spectra for four-variable combinations are uninformative and not shown for clarity. The black bars identify the peaks obtained by both the PAI-based AMBIENCE search algorithm and EXS methods, the white bars indicate the peaks obtained by the EXS alone, and the gray bars indicate the peaks obtained by AMBIENCE alone.

The KWII spectrum in Fig. 2.2B demonstrates that the AMBIENCE search algorithm detects all peaks with significant GEI without the enumeration of all possible combinations that is required in the EXS approach. The KWII values of combinations containing only informative variables are unaffected by the LD between uninformative variables. The KWII spectrum correctly identifies the one-variable-containing peaks that demonstrate the critical roles of $E1$, $E2$, SNP 1, and SNP 2 variables in the underlying model. A strong peak corresponding to the $\{1,2,C\}$ interaction is also identified. These peaks are also featured in the KWII spectrum of EXS (as indicated by the gray bars). None of the significant peaks involving an interaction between the known interacting variables is omitted in the spectrum of the AMBIENCE search algorithm. All the peaks present in the KWII spectrum that were detected by the EXS method only (white bars in Fig. 2.2B) have very low magnitudes compared to the stronger peaks with known interactions. These results demonstrate that the KWII correctly identifies all known GEIs in the case study. Notably, the $\{E1,E2,C\}$ combination was not present among the top 20 two-variable combinations with the highest KWII values in either the AMBIENCE and EXS methods, denoting the absence of any interaction between $E1$ and $E2$.

2.4.1.1 PAI removes the effects of LD on TCI

The TCI spectrum of Case Study 1 (Fig. 2.2C) shows prominent peak changes relative to Fig. 2.2D for combinations $\{2,3,4,C\}$, $\{1,3,4,C\}$, $\{E4,3,4,C\}$, $\{E2,3,4,C\}$, $\{E1,3,4,C\}$, and $\{3,4,C\}$ that are caused by the LD between the uninformative variables, SNP 3 and SNP 4. These peak changes are absent in the PAI spectrum of Case Study 1 (Fig. 2.2D), demonstrating that the PAI is unaffected in the presence of LD between uninformative SNP variables. Thus, the PAI is more effective than the TCI in detecting GEI when LD between uninformative SNP variables is present.

Figures 2.3A and 2.3B systematically examine the effect of LD between two SNPs on the TCI and PAI using Case Study 1. In Fig. 2.3A we varied the LD between SNP 3 and SNP 4 (both were not associated with the disease phenotype) from 0 to 1. The representative combinations $\{3,4,C\}$ and $\{E1,3,4,C\}$ are presented because they include the two SNP variables in LD with each other (SNP 3 and SNP 4), the risk-increasing environmental variable $E1$, and the phenotype variable. The results (Fig. 2.3A) demonstrate that, as expected, the increasing LD between SNP 3 and SNP 4 contributes to the TCI of the combinations containing these variables. In contrast to the TCI (which increases with increasing LD), the PAI remains unchanged: $0.0055 \pm$ SD 0.0027 in the absence of LD and 0.0043 ± 0.0022 for LD $= 0.9$. Importantly, for the $\{3,4,C\}$ combination, the PAI remains at a value close to zero, indicating the correct detection of no association between the SNPs and the disease status despite the high LD. In the absence of LD, the TCI also correctly assessed the lack of association between SNP 3 and SNP 4 with the phenotype (0.0086 ± 0.0032); however, when LD between SNP 3 and SNP 4 was increased to 0.9, the TCI increased to a value of 1.19 ± 0.037. In the second combination assessed, we include a risk-increasing environmental variable, $E1$, to $\{3,4,C\}$. Because the $\{E1,3,4,C\}$ combination contains the disease-associated $E1$ environmental variable, we correctly anticipated that both PAI and TCI values would be larger for $\{E1,3,4,C\}$ than $\{3,4,C\}$. In the absence of LD, the PAI and TCI were 0.170 ± 0.021 and 0.178 ± 0.021, respectively. When the LD between SNP 3 and SNP 4 was increased to 0.9, the TCI value combination increased more than tenfold whereas the PAI was constant at 0.170 ± 0.019. The PAI retained the disease association due to the presence of $E1$, while remaining unaffected by the LD between SNP 3 and SNP 4. This example highlights the validity of the statement made earlier that in the presence of LD, the PAI is a more effective metric than the TCI for detecting disease phenotype–associated GEI.

In the next set of experiments, we again examined the effect of LD between two SNPs, with the exception that one of the SNPs was associated with the disease phenotype. We modified Case Study 1 by introducing LD between SNP 3 (which is not associated with the phenotype) and SNP 2 (which is involved in the phenotype). For this case (Fig. 2.3B), the representative combinations $\{2,3,C\}$ and $\{E1,2,3,C\}$ are presented. In the presence of LD $= 0.9$, the TCI values of the $\{2,3,C\}$ and $\{E1,2,3,C\}$ combinations increased rapidly. In contrast, the PAI values of the $\{2,3,C\}$ and $\{E1,2,3,C\}$ combinations remained constant. Again, the results clearly indicate that the PAI effectively captured the genetic-risk-increasing and environmental-risk-increasing information in the data, while simultaneously eliminating the spurious effects of LD in the $\{2,3,C\}$ and $\{E1,2,3,C\}$ combinations.

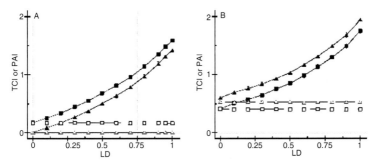

Figure 2.3 (A) TCI and PAI for various levels of LD between SNP 3 and SNP 4 (both are uninformative SNPs) for Case Study 1. (B) Effects of LD between the informative SNP 2 and the uninformative SNP 3 when Case Study 1 was modified by introducing LD between SNP 2 and SNP 3. In (A), the combinations $\{3,4,C\}$ and $\{E1,3,4,C\}$ are shown in triangles and squares, respectively. In (B), the combinations $\{2,3,C\}$ and $\{E1,2,3,C\}$ are shown in triangles and squares, respectively. In both figures, the filled symbols are the TCI, whereas the open symbols are the PAI. Reprinted with permission from Ref. [18], Copyright 2008 by the Genetics Society of America.

2.5 Algorithms

The major challenges in GEI analysis are the high dimensionality of the data sets and the combinatorial explosion in the number of possible interactions. The addition of extra dimensions to a mathematical space exponentially increases the hypervolume in which the data are distributed (part of the so-called *curse of dimensionality*), which increases the computational complexity and degrades the statistical power of algorithms. However, it is the combinatorial explosion that is more computationally problematic.

The relationship between the number of interactions and the number of predictors increases with extraordinary rapidity because there are $^{n}C_{k}$ (also known as the binomial coefficient or n-choose-k function) ways of selecting a subset of k attributes for assessing interactions among n attributes. This combinatorial growth makes it computationally difficult, if not impossible, to exhaustively search the full range of genetic and environmental variables for potential interactions associated with diseases in epidemiologic studies. Although the increasing trend in the number of combinations due to an increase in the number of variables is well known, the rapidity of this growth is often underestimated. This rapid growth necessitates the use of effective search methods to identify the most promising interactions.

The principal problems in GEI analysis are:

- Identifying more effective metrics to characterize and detect GEI associated with the risk of disease.
- Developing efficient algorithms that can tackle combinatorial complexity sufficiently to highlight regions of the combinatorial space containing the most promising interactions.
- Identifying efficient methods for statistical significance assessments. Permutation-based statistical assessments can be time consuming for a large number of interactions as a large number of permutations have to be done independently for each combination. Multiple testing is an inherent challenge in GEI analyses.
- Implementing modeling strategies to enable the user to identify the critical GEI from among the limited, but still numerous, combinations identified. A challenge is that the structure of the underlying dependencies is not known. Nonparametric rather than parametric modeling approaches are preferred.

2.5.1 Noninformation Theoretic Methods

Regression methods [22] such as logistic and log-linear regression have been investigated extensively for GGI and GEI analysis. These methods use a linear or linearized framework, and while interactions identified by such methods can be interpreted using the structure and parameterization of the (statistical) model, they do not capture even the simplest physicochemical descriptions of noncovalent biological interactions. Additionally, complete parameterization of a kth order interaction in regression models results in 2^k parameters for a genotype with three states. The large number of parameters in the output makes interpretation from large data set analyses difficult [23]. Thus, the benefit of interpreting interactions within the framework of a statistical model comes with steep costs in both computational complexity and detection potential. In contrast, the KWII provides a single, flexible, and interpretable parameter [23].

The multifactor dimensionality reduction (MDR) technique (and software) for identifying and analyzing GEI was developed by Ritchie and her colleagues [3, 24, 25]. MDR is based on nonparametric multifactor models and allows statistical and cross-validation analysis of GGI and GEI for balanced case-control and discordant sib-pair designs [24–27]. MDR uses constructive induction wherein the dimensionality of the multi-locus genotype is systematically reduced by pooling into high- and low-risk groups [3]. It then selects the combinations that maximize the accuracy of predicting cases and controls by cross validation. MDR, like entropy-based approaches, has the attractive property that it is nonparametric and does not require model specification to search for interactions. However, at its heart, this algorithm is an EXS with a high computational burden and difficulty handling large data sets [28]; this burden is exacerbated by the cross validation within runs. The MDR method has been extended to unbalanced data sets [29], and a theoretical analysis of MDR has shown the similarity of the MDR classifier to the naive Bayes classifier [30]. The MDR approach has been used to study GEI in atrial fibrillation, autism, and diabetes mellitus [31–34]. MDR is applicable only to discrete phenotypes, not to quantitative traits.

There are only a few methods for GEI and GGI analysis of quantitative traits. The combinatorial partitioning method (CPM), which shares similarities with MDR, is an approach to identify multilocus genotypes capable of predicting quantitative trait levels [35]. The CPM is computationally very intensive, and Culverhouse, Klein, and Shannon (2004) advocated the restricted partition method (RPM) [36]. Although the RPM is computationally more efficient than the CPM, it still requires significant computational effort for high-dimensional data. The RPM is strongly dependent on good partitioning of genotypes into subgroups by iterative application of multiple comparison tests [37]. Both the CPM and the RPM do not allow adjustment for covariates. The recently proposed generalized MDR (GMDR) method employs the generalized linear model (GLM) framework for scoring in conjunction with MDR for dimensionality reduction. GMDR enables covariate corrections and handles both discrete and continuous phenotypes in population-based study designs. GMDR employs the same risk-pooling (dimensionality reduction) strategy as MDR and yields the original MDR as a special case when covariates are not present and the phenotypes are discrete [38]. However, despite the availability of a more efficient parallel computing implementation [26], MDR and its variants (including GMDR) are computationally intensive, especially when more than 10 polymorphisms need to be evaluated [25].

The pedigree disequilibrium test (PDT) [39] and its extensions are one of the few parametric approaches for GGI analysis with available software (UNPHASED v3.10) [40]. UNPHASED can be used efficiently for case/parent trios, affected sib-pairs, and case-control study designs; additionally, either *cis-* or *trans*-phase (statistical) GGI can be detected. The PDT approach has been used to extend the MDR, which was initially limited in its capacity to include potentially informative family data beyond single matched pairs in each family, to family-based study designs [41]. Together, these approaches allow for tests of GGI in family data to be constructed, while simultaneously reducing dimensionality of the data.

2.5.2 Information Theoretic Algorithms

Broadly viewed, regression trees and random forests rely on information gain and are within the family of entropy-based algorithms. Information gain is identical to a reduction in entropy. As they are interested in providing a model based on rules that have yes or no answers, regression/decision trees focus on reducing the entropy in a system due to specific observations of a random variable. Random forests are an extension of decision trees. In a random forest analysis trees are "grown" from a training set of the data and a random vector of predictors (from the data) to create a "forest": the most popular predictors [42] in the forest are then selected. This approach operates under the assumption that if an interaction is present, it will be present no matter how you partition the data; however, standard tree-based approaches can be sensitive to small changes in the training partitions of the data [43]. Modifications to the random forest approach have enabled their utilization in interaction detection on genome-wide data sets [44–46]. The results from these efforts are mixed; however, their effectiveness in the presence of genetic complexities such as LD and heterogeneity have not yet been systematically investigated or considered.

Another algorithm that has shown promising results for large data sets is BOOST [28]. It is an efficient algorithm that is powered by the KLD and has been applied to seven genome-scale data sets from the Wellcome Trust Case-Control Consortium (WTCCC). This algorithm starts with a log-linear (likelihood) model approach and builds up to a KLD definition by removing the likelihood of the homogenous association model from the saturated model, leaving only the contribution from the interaction term. This methodology highlights the known existence of a close relationship between entropy-based approaches and log-likelihood approaches [23]. Despite this relationship, no closed-form solution exists for the homogenous association model, and the BOOST algorithm requires using the Kirkwood superposition approximation (KSA) as a surrogate for this value. Interestingly, the authors failed to note the relationship between the KWII and the KSA, which has been known since the work of Bratko and Jakulin [47, 48]. One potential shortcoming of the BOOST algorithm is how it will perform in the presence of genetic heterogeneity, which is frequent in data sets from complex diseases [49]. Although BOOST has not been examined in the presence of heterogeneity, it is likely that the utilization of the homogenous association model to remove lower-order effects would have negative consequences on its power. Further, the KWII accomplishes the same task of quantitating only the information that is gained from accounting for all variables of interest, simultaneously [11], without the need for approximation statistics.

AMBIENCE uses the KWII to quantitate the presence of an interaction, while using the PAI to search for informative combinations, as the computational burden of the KWII is much greater than the PAI. AMBIENCE exploits the monotonic nature of the PAI to guide a directed, hill-climbing, greedy search that can efficiently identify the most promising GEI. Unlike MDR, AMBIENCE and BOOST are not exhaustive; they both reduce the number of calculations of subsequent orders by selecting a prespecified number of combinations to carry ahead.

The KWII has been shown to outperform MDR on large data sets [18]. Although both the PAI and KWII are sensitive to pure epistasis, MDR is likely to have greater power to detect pure epistasis as

AMBIENCE will not carry forward SNPs without at least marginal main effects. AMBIENCE makes no assumption of homogeneity and is one of the few methods that have been shown to perform well in the presence of both LD and statistical heterogeneity [18, 49].

There are some fundamental differences and unique advantages to AMBIENCE compared to the widely used MDR, GMDR, and PDT approaches. MDR, GMDR, and PDT do different things and ask different questions. For example, PDT is particularly useful for family-based study designs and can accommodate missing data, whereas MDR is a nonparametric method for case-control study designs, and GMDR is capable of handling continuous covariates. These available methods are computationally prohibitive for analyzing interactions in genome-wide data.

In interaction analysis, *model synthesis* can be defined as the process of identifying a parsimonious set of variable combinations capable of explaining the phenotype. The variable combinations (and their KWIIs) identified in the AMBIENCE output are the input for the model synthesis procedure AMBROSIA. Model synthesis in AMBROSIA is based on the KWII; its key advantage is that repeated refitting to the data is *not* necessary. The KWII can be leveraged for model synthesis because of the relationship of the KWII to the KLD between the joint probability density and the model constructed using all pairwise dependencies [15]; additionally, the KLD is the expected log-likelihood [10]. To avoid over- and underfitted models, the corrected, or small-sample, Akaike information criterion (abbreviated for simplicity as AIC) is used for model selection [50]. The AIC imposes a penalty on the log-maximum-likelihood function for increasing the number of fitted free parameters in a model [51, 52]. AMBROSIA can be feasibly implemented using model synthesis procedures that nominally resemble forward selection and backward elimination methods in regression.

The Möbius transform is a powerful tool that can enable efficient computation of the KWII, particularly when the KWII of all subsets is needed. A fast Möbius transform for computing the KWII of all subsets was developed by Thoma [53] and can be modified to yield an algorithm that provides subsets of order k or lower [23]. Tritchler et al. proposed an algorithm, *Möby Quick*, based on the fast Mobius transform for GEI analysis. Williams [54] utilized a Möbius transform to obtain a nonnegative multivariate interaction information. Approaches based on the fast Möbius transform require more investigation in GEI algorithms.

2.6 Applications of Information Theory to Interaction Analysis

Here we provide a selective rather than an exhaustive overview of research that has used information theory for interaction analysis.

Several reports have used the KLD for genetic analysis [6, 55–57]. The KLD is a measure of the distance between two distributions because it measures the inefficiency of assuming that the distribution is q when the true distribution is p. In genetic analyses, the most frequent application of the KLD has been for two-group comparisons such as those used to evaluate ancestry informative markers [55–57]. However, the KLD has also been proposed as a multilocus LD measure to enable identification of TagSNPs [6], and our group has adapted the KLD for analytical visualization [4, 5]. Information theory statistics employing entropy-based principles have been proposed for genome-wide data analysis to test for allelic association with a phenotype [58–60]. Entropy-based methods for two-locus interactions have also been proposed recently and were found to confirm the negative epistasis between sickle cell anemia and alpha-thalassemia genetic variations against malaria [61].

In our research with the KWII, we have built on the work of Jakulin and Bratko extensively [14, 15, 47, 48]. We also have developed visual analysis methods based on the KWII or PAI spectra, which are graphical summaries of KWII and PAI results that facilitate interpretation of informative genetic variations, environmental variables, GEI, and GGI. Jakulin and Bratko [14, 15, 47, 48] proposed visualization approaches that included information graphs and interaction dendrograms. In information graphs, the graphical representation of nodes includes information regarding the percentage of the entropy reduction from the class variable, while the edge representation includes information on whether the interaction is synergistic or redundant. Interaction dendrograms are obtained by hierarchical clustering on a distance matrix derived from the magnitude of the interaction information [3]. Moore incorporated interaction dendrograms into MDR [3], and it has been used to study the effects of SNP genotypes and smoking on bladder cancer risk [62].

Calle et al. [63] analyzed case-control data from the Spanish bladder cancer study (SBCS) and used an entropy-based approach as the first of a two-part detection methodology to overcome the computational burdens of MDR. The data set contained 282 SNPs and represented 108 genes in the inflammatory pathway from 2627 patients (1356 cases and 1271 controls). By using mutual information to select potentially informative SNPs to carry forward into their model-based MDR (MB-MDR) algorithm, significant second- and third-order interactions were found for bladder cancer patients who were currently smokers. Their search metric for selecting these SNPs was synergy (SYN), which they defined for second-order interactions as

$$\text{SYN}(G_1, G_2; Y) = I(G_1, G_2; Y) - [I(G_1; Y) + I(G_2; Y)]$$

where G_1 and G_2 are two SNPs, Y is the (binary) phenotype, and $I(v;Y)$ represents the mutual information between the predictor(s) in the set v and the phenotype, Y. This metric is presented intuitively as the amount of information in the full interaction that is not associated with lower-order interactions; however, it can easily be shown that for second-order interactions, their definition of synergy is equal to the negative multiple mutual information of the set, such that

$$\text{SYN}(G_1, G_2; Y) = -I(G_1; G_2; Y)$$

For third-order (and higher) interactions, however, a different definition of synergy is defined, such that

$$\text{SYN}(G_1, G_2, \ldots, G_n; Y) = I(G_1, G_2, \ldots, G_n; C) - \max_{\substack{\text{all partitions} \\ \{S_j\} \text{ of } S}} \sum_j I(S_j; C)$$

The higher-order synergy is interpreted as the additional amount of information that is provided by the entire combination after removing what can 'best' be described from its parts. Although they have multiple definitions of synergy depending on the order of the interaction, the authors were able to use entropy-based search metrics to identify predictors that could be passed on to the more computationally intensive MB-MDR on a data set that may not have been able to be analyzed by this method previously. The authors do not provide SNP identifiers, biological interpretation, or comparison of the results, but simply present their work as an improvement to existing methodologies for interaction analysis and of large case-control data sets.

Shervais et al. [64] applied reconstructability analysis to type 2 non-insulin-dependent diabetes (NIDDM) case-control data. Reconstructability analysis is an information-theoretic modeling strategy that constructs a lattice of projections, which is essentially a decomposition of the saturated model (the full data) into all possible lower-order models of association and interaction [65]. The fit of the reduced models (based on the marginal distributions) to the data is assessed using a maximum-entropy method [23]. Compiling previous results from Cox [66], Horikawa [67], and Tsalenko [68], who identified regions on chromosome 2 (chr2) and chromosome 15 (chr15) associated with NIDDM, Shervais et al. filtered a set of 202 SNPs from 220 subjects. After filtering, they were left with 63 SNPs on chr2, and 35 SNPs on chr15. Thirteen of the top 16 chr15 SNPs identified by Shervais et al. were also identified by Horikawa [67], while 10 of Shervais et al.'s top 15 chr2 SNPs were concordant with those from Tsalenko [68]. Their epistasis analysis captured multiple interacting SNPs between the regions identified in the original studies, while also detecting an interaction between the chr2 *GPR35/CAPN10* region and the chr15 *HNF6* gene, which has been implicated in a form of diabetes called *maturity-onset diabetes of the young* [64].

Kasturi, Acharya, and Ramanathan provided one of the earlier demonstrations of information theory for clustering gene expression data [69]. Temporal gene expression patterns were converted directly to probability distributions by normalization with the area under the curve. The KLD was used as a pairwise distance measure for clustering. The clusters identified with the information theoretic metric were found to be superior to hierarchical clustering with statistical metrics [69].

Hernandez-Lemus et al. [70] analyzed biological expression pseudo-time-series data by extending methods from linguistic analysis. Using genome-wide expression data from 67 papillary thyroid cancer (PTC) patients on Affymetrix HGU133Plus2 chips, the authors collected gene expression data from patients at different stages of malignancy, and subsequently ordered the data based on galectin-3 expression of the patients—this was used a surrogate of disease progression. This pseudo time series was intended to mimic the dynamics of cancer progression. The approach behind this work [70] was to consider gene regulatory networks (GRNs) as consisting of communication channels for cellular processes. In the first step, the continuous expression data were transformed to binary "words." Information theory was used to define the joint probability density (JPD) for the expression of each gene and then used to calculate a pairwise distance measure called the information-based similarity (IBS) between each of the ordered vectors, which contained the normalized likelihood of the word, weighted by the Shannon entropy [70]. The IBS was successfully used to detect second-order interactions in a complex dynamic environment containing non-linear dependencies. Although this approach could be used for higher-order interactions, the sample size would need to approach 10^3 [70] which is unrealistic for current genomic studies. This elegant application of information theory yielded networks supported by other studies of thyroid function, including roles for beta 1-catenin, retinoid X receptor gamma transcription enhancer, and cytokeratin-19. Additionally of note was a cluster of genes formed by *SENP6, NBPF11, ZNF611, ITPR2, USP34,* and *ZBTB2D.* The authors highlighted the finding that these genes (with the exception of *ITPR2*) are not usually associated with cancer but are important genes in thyroid malfunction [71]. Additionally, *USP34* has been confirmed as overexpressed in both mRNA and proteomic analyses of adrenal thyroid neoplasms [72].

2.7 Critiques of Information Theory Approaches to Interaction Analysis

A common criticism to the KWII for three or more variables is the supposedly undesirable property that it can take on values that are positive, negative, or zero. In contrast, the Shannon entropy for quantitating information [7] is always positive (or zero), and for this reason, information theory purists expect every information metric be positive. In the case of two variables, the KWII is identical to the mutual information, which is always positive (or zero) [73]. However, for three variables, a relationship between two of the variables may be disrupted by the third, which may lead to a loss of information that was gained from the original two variables: this would result in a negative KWII. Most commonly, the interpretation of this negative KWII, or interaction information, is redundancy amongst the variables [54, 73, 74]. Bell [75] stated, "It is not quite clear yet what it means to have a negative co-information, but it is clear that a non-zero value signals the existence of kth-order dependency." Although it has been shown that KWII is a mathematically sound, and useful, metric [73, 75, 76], much of the debate (particularly among systems theorists) is focused on the semantics of whether a metric such as the KWII, which can be negative, should be called an information metric.

The underlying reasons for why the KWII can be negative have been well understood for some time [14, 18, 77]. The KWII is related intimately to the KSA [78], which decomposes lower-order dependencies usefully but does not guarantee a normalized probability density function. The perhaps narrow focus on why the KWII can be negative and cannot strictly be referred to as a Shannon-type information theory metric fails to recognize the strengths of the KWII in searching large combinatorial landscapes and in identifying potentially useful predictors. Indeed continuous entropy, which is well entrenched in information theory, does not share the same properties as Shannon entropy.

Krippendorf showed that the KWII at higher orders can be viewed as the difference between Shannon-type information, and redundancy [79]. Leydesdorff [80] highlighted the usefulness of interaction information and demonstrated that a value of zero for the KWII simply implies that the information is offset by the redundancy. A true Shannon-type information multivariate metric can be obtainable through iterative methods [79]. Additionally, nonnegative decompositions of multivariate mutual information have been developed using a redundancy lattice approach [54]. Iterative methods present a computational burden for very large data sets [81], and although advancements in computer speed could make this less important [79], the issue of scalability to genome-scale data has not been demonstrated.

There is a need for further investigation into minimizing the effects of study design on interaction information as they are not invariant to case-control sampling [82] under certain circumstances. However, the usefulness of the KWII for analyzing individual data sets is largely unaffected because the special cases of invariance will not affect inference and because the magnitude of the KWII is not interpreted or used in the same way as model-dependent metrics such as the odds ratio [23].

Another criticism of information theoretic methods is that there is no formal a priori genetic mechanism or particular linear (or nonlinear) mathematical form used for GEI identification. This is in direct contrast to regression methods, GMDR, and the PDT, which are refined for family-based studies. The absence of assumptions should be viewed as one of the strengths of information theory methods because it makes the method nonparametric and enables the detection of complex linear and nonlinear multivariate relationships, simultaneously, in the data. Additionally, the freedom from defining a mechanism a priori minimizes the need for user input.

The KWII and the PAI have excellent power for detecting pure epistasis when it occurs. However, greedy search algorithms such as AMBIENCE depend on marginal effects and could, in theory, miss pure epistatic interactions. This can be easily remedied with strategies that selectively sample two-order combinations [18]. However, for applications in statistical genetics and pharmacogenetics, purely epistatic interactions are extremely rare [17, 18, 49]. Further, simulations containing pure epistasis (third-order interactions) require artificially contrived assumptions and readily produce marginal effects upon minor perturbation with allele frequency and LD changes.

2.8 Conclusions

In conclusion, this chapter focused on the principles and applications of information theoretic methods in gene–gene and gene–environment interaction analysis. Information theoretic methods complement, and potentially extend, the capabilities of existing statistical genetics methods. The salient biological and computational advantages are as follows:

- The information theoretic metrics have mathematically optimal properties with rich theoretical foundations in hypothesis testing, modeling, and information theory [9, 10].

- The PAI and the KWII are capable of detecting both linear and nonlinear dependencies. They can detect GEI regardless of the presence or absence of main effects.
- Traditional LD measures are pairwise in character and are typically computed for contiguous SNPs. The PAI and the KWII are multivariate in character and assess the joint dependence between variables generalized to multiple, nonadjacent genetic variations [6].
- The approach can be used when the genetic and environmental variables have different numbers of classes or when the phenotype has more than two classes (or is continuous).
- These metrics provide a basis for efficient interaction identification and interaction-modeling algorithms: AMBIENCE, an algorithm that efficiently reduces computational complexity from combinatorial/exponential time to polynomial time [18], and AMBROSIA, a model synthesis method that does not require repeated refitting of the data, have been demonstrated [83].
- The approach is novel and versatile: It is effective not just for discrete phenotypes but has also been extended to quantitative traits, as well as rate/count data [17, 19].

References

1. Manolio, T. A. (2010). Genomewide association studies and assessment of the risk of disease. *N. Engl. J. Med.*, **363**(2), pp. 166–176.

2. Goldstein, D. B. (2009). Common genetic variation and human traits. *N. Engl. J. Med.*, **360**(17), pp. 1696–1968.

3. Moore, J. H., Gilbert, J. C., Tsai, C. T., Chiang, F. T., Holden, T., Barney, N., and White, B. C. (2006). A flexible computational framework for detecting, characterizing, and interpreting statistical patterns of epistasis in genetic studies of human disease susceptibility. *J. Theor. Biol.*, **241**(2), pp. 252–261.

4. Bhasi, K., Zhang, L., Brazeau, D., Zhang, A., and Ramanathan, M. (2006). Information-theoretic identification of predictive SNPs and supervised visualization of genome-wide association studies. *Nucleic Acids Res.*, **34**(14), p. e101.

5. Bhasi, K., Zhang, L., Brazeau, D., Zhang, A., and Ramanathan, M. (2006). VizStruct for visualization of genome-wide SNP analyses. *Bioinformatics*, **22**(13), pp. 1569–1576.

6. Liu, Z. and Lin, S. (2005). Multilocus LD measure and tagging SNP selection with generalized mutual information. *Genet. Epidemiol.*, **29**(4), pp. 353–364.

7. Shannon, C. E. (1997). The mathematical theory of communication (reprinted). *M. D. Computing*, **14**(4), pp. 306–317.

8. McGill, W. J. (1954). Multivariate information transmission. *Psychometrika*, **19**, pp. 97–116.

9. Cover, T. M. and Thomas, J. A. (1991). *Elements of Information Theory*, p. 542 (Wiley Series in Telecommunications, Wiley, New York).

10. Haykin, S. (1999). *Neural Networks: A Comprehensive Foundation* (College, New York).

11. Chanda, P., Zhang, A., Brazeau, D., Sucheston, L., Freudenheim, J. L., Ambrosone, C., and Ramanathan, M. (2007). Information-theoretic metrics for visualizing gene-environment interactions. *Am. J. Hum. Genet.*, **81**(5), pp. 939–963.

12. Han, T. (1980). Multiple mutual informations and multiple interactions in frequency data. *Inf. Control*, **46**(1), pp. 26–45.

13. Fano, R. M. (1961). *Transmission of Information: A Statistical Theory of Communications* (MIT Press, Cambridge, MA).

14. Jakulin, A. (2005). Machine learning based on attribute interactions. In *Computer Science*, p. 240 (University of Ljubljana, Ljubljana, Slovenia).

15. Jakulin, A. and Bratko, I. (2004). Testing the significance of attribute interactions. In *Proceedings of the Twenty-First International Conference on Machine Learning (ICML-2004)* (Banff, Canada).

16. Bell, A. J. (2002). Co-information lattice. In *4th International Symposium on Independent Component Analysis and Blind Source Separation* (Nara, Japan).

17. Chanda, P., Sucheston, L., Liu, S., Zhang, A., and Ramanathan, M. (2009). Information-theoretic gene-gene and gene-environment interaction analysis of quantitative traits. *BMC Genomics*, **10**, p. 509.

18. Chanda, P., Sucheston, L., Zhang, A., Brazeau, D., Freudenheim, J. L., Ambrosone, C., and Ramanathan, M. (2008). AMBIENCE: a novel approach and efficient algorithm for identifying informative genetic and environmental associations with complex phenotypes. *Genetics*, **180**(2), pp. 1191–1210.

19. Knights, J. and Ramanathan, M. (2012). An information theory analysis of gene-environmental interactions in count/rate data. *Hum. Hered.*, **73**(3), pp. 123–38.

20. Watanabe, S. (1960). Information theoretical analysis of multivariate correlation. *IBM J. Res. Dev.*, **4**, pp. 66–82.

21. Fisher, R. (1936). The user of multiple measurements in axonomic problems. *Ann. Eugenics*, **7**(17), pp. 179–188.

22. Cordell, H. J. (2002). Epistasis: what it means, what it doesn't mean, and statistical methods to detect it in humans. *Hum. Mol. Genet.*, **11**(20), pp. 2463–2468.

23. Tritchler, D. L., Sucheston, L., Chanda, P., and Ramanathan, M. (2011). Information metrics in genetic epidemiology. *Stat. Appl. Genet. Mol. Biol.*, **10**(1), Article 12.

24. Hahn, L. W., Ritchie, M. D., and Moore, J. H. (2003). Multifactor dimensionality reduction software for detecting gene-gene and gene-environment interactions. *Bioinformatics*, **19**(3), pp. 376–382.

25. Ritchie, M. D., Hahn, L. W., Roodi, N., Bailey, L. R., Dupont, W. D., Parl, F. F., and Moore, J. H. (2001). Multifactor-dimensionality reduction reveals high-order interactions among estrogen-metabolism genes in sporadic breast cancer. *Am. J. Hum. Genet.*, **69**(1), pp. 138–147.

26. Bush, W. S., Dudek, S. M., and Ritchie, M. D. (2006). Parallel multifactor dimensionality reduction: a tool for the large-scale analysis of gene-gene interactions. *Bioinformatics*, **22**(17), pp. 2173–2174.

27. Ritchie, M. D., Hahn, L. W., and Moore, J. H. (2003). Power of multifactor dimensionality reduction for detecting gene-gene interactions in the presence of genotyping error, missing data, phenocopy, and genetic heterogeneity. *Genet. Epidemiol.*, **24**(2), pp. 150–157.

28. Wan, X., Yang, C., Yang, Q., Xue, H., Fan, X., Tang, N. L., and Yu, W. (2010). BOOST: a fast approach to detecting gene-gene interactions in genome-wide case-control studies. *Am. J. Hum. Genet.*, **87**(3), pp. 325–340.

29. Velez, D. R., White, B. C., Motsinger, A. A., Bush, W. S., Ritchie, M. D., Williams, S. M., and Moore, J. H. (2007). A balanced accuracy function for epistasis modeling in imbalanced datasets using multifactor dimensionality reduction. *Genet. Epidemiol.*, **31**(4), pp. 306–315.

30. Hahn, L. W. and Moore, J. H. (2004). Ideal discrimination of discrete clinical endpoints using multilocus genotypes. *In Silico Biol.*, **4**(2), pp. 183–194.

31. Tsai, C. T., Lai, L. P., Lin, J. L., Chiang, F. T., Hwang, J. J., Ritchie, M. D., Moore, J. H., Hsu, K. L., Tseng, C. D., Liau, C. S., and Tseng, Y. Z. (2004).

Renin-angiotensin system gene polymorphisms and atrial fibrillation. *Circulation*, **109**(13), pp. 1640–1646.

32. Motsinger, A. A., Donahue, B. S., Brown, N. J., Roden, D. M., and Ritchie, M. D. (2006). Risk factor interactions and genetic effects associated with post-operative atrial fibrillation. *Pac. Symp. Biocomput.*, pp. 584–595.

33. Ma, D. Q., Whitehead, P. L., Menold, M. M., Martin, E. R., Ashley-Koch, A. E., Mei, H., Ritchie, M. D., Delong, G. R., Abramson, R. K., Wright, H. H., Cuccaro, M. L., Hussman, J. P., Gilbert, J. R., and Pericak-Vance, M. A. (2005). Identification of significant association and gene-gene interaction of GABA receptor subunit genes in autism. *Am. J. Hum. Genet.*, **77**(3), pp. 377–388.

34. Cho, Y. M., Ritchie, M. D., Moore, J. H., Park, J. Y., Lee, K. U., Shin, H. D., Lee, H. K., and Park, K. S. (2004). Multifactor-dimensionality reduction shows a two-locus interaction associated with Type 2 diabetes mellitus. *Diabetologia*, **47**(3), pp. 549–554.

35. Nelson, M. R., Kardia, S. L., Ferrell, R. E., and Sing, C. F. (2001). A combinatorial partitioning method to identify multilocus genotypic partitions that predict quantitative trait variation. *Genome Res.*, **11**(3), pp. 458–470.

36. Culverhouse, R., Klein, T., and Shannon, W. (2004). Detecting epistatic interactions contributing to quantitative traits. *Genet. Epidemiol.*, **27**(2), pp. 141–152.

37. Heidema, A. G., Boer, J. M. A., Nagelkerke, N., Mariman, E. C. M., van der A, D. L., and Feskens, E. J. M. (2006). The challenge for genetic epidemiologists: how to analyze large numbers of SNPs in relation to complex diseases. *BMC Genet.*, **7**, p. 23.

38. Lou, X. Y., Chen, G. B., Yan, L., Ma, J. Z., Zhu, J., Elston, R. C., and Li, M. D. (2007). A generalized combinatorial approach for detecting gene-by-gene and gene-by-environment interactions with application to nicotine dependence. *Am. J. Hum. Genet.*, **80**(6), pp. 1125–1137.

39. Martin, E. R., Monks, S. A., Warren, L. L., and Kaplan, N. L. (2000). A test for linkage and association in general pedigrees: the pedigree disequilibrium test. *Am. J. Hum. Genet.*, **67**(1), pp. 146–154.

40. Dudbridge, F. (2003). Pedigree disequilibrium tests for multilocus haplotypes. *Genet. Epidemiol.*, **25**(2), pp. 115–121.

41. Martin, E. R., Ritchie, M. D., Hahn, L., Kang, S., and Moore, J. H. (2006). A novel method to identify gene-gene effects in nuclear families: the MDR-PDT. *Genet. Epidemiol.*, **30**(2), pp. 111–123.

42. Breiman, L. (2001). Random forests. *Mach. Learn.*, **45**(1), pp. 5–32.

43. Elith, J., Leathwick, J. R., and Hastie, T. (2008). A working guide to boosted regression trees. *J. Animal Ecol.*, **77**(4), pp. 802–813.

44. Garcia-Magarinos, M., López-de-Ullibarri, I., Cao, R., and Salas, A. (2009). Evaluating the ability of tree-based methods and logistic regression for the detection of SNP-SNP interaction. *Ann. Hum. Genet.*, **73**(Pt 3), pp. 360–369.

45. Sun, Y. V., Cai, Z., Desai, K., Lawrance, R., Leff, R., Jawaid, A., Kardia, S. L. R., and Yang, H. (2007). Classification of rheumatoid arthritis status with candidate gene and genome-wide single-nucleotide polymorphisms using random forests. *BMC Proc.*, **1**(Suppl 1), p. S62.

46. Kim, Y., Wojciechowski, R., Sung, H., Mathias, R. A., Wang, L., Klein, A. P., Lenroot, R. K., Malley, J., and Bailey-Wilson, J. E. (2009). Evaluation of random forests performance for genome-wide association studies in the presence of interaction effects. *BMC Proc.*, **3**(Suppl 7), p. S64.

47. Jakulin, A. and Bratko, I. (2003). Analyzing attribute dependencies. In *Proceedings of the 7th European Conference on Principles and Practice of Knowledge Discovery in Databases (PKDD 2003)* (Springer, Cavtat-Dubrovnik, Croatia).

48. Jakulin, A., Bratko, I., Smrke, D., Demar, J., and Zupan, B. (2003). Attribute interactions in medical data analysis. In *Proceedings of the 9th Conference on Artificial Intelligence in Medicine in Europe (AIME 2003)* (Springer, Protaras, Cyprus).

49. Sucheston, L., Chanda, P., Zhang, A., Tritchler, D., and Ramanathan, M. (2010). Comparison of information-theoretic to statistical methods for gene-gene interactions in the presence of genetic heterogeneity. *BMC Genomics*, **11**, p. 487.

50. Hurvich, C. M. and Tsai, C. L. (1995). Model selection for extended quasi-likelihood models in small samples. *Biometrics*, **51**(3), pp. 1077–1084.

51. Akaike, H. (1974). A new look at the statistical model identification. *IEEE Trans. Automat. Contr. AC*, **19**, pp. 716–723.

52. Hurvich, C. M. and Tsai, C. L. (1989). Regression and time series model selection in small samples. *Biometrika*, **76**, pp. 297–293.

53. Thoma, H. M. (1991). Belief function computations. In *Conditional Logic in Expert Systems*, Vol. I, pp. 269–308, R. Goodman, et al. (eds.) (Elsevier Science, Amsterdam, North-Holland).

54. Williams, P. (2010). *Nonnegative Decomposition of Multivariate Information* (Indiana University), http://www.arXiv.org.

55. Rosenberg, N. A., Li, L. M., Ward, R., and Pritchard, J. K. (2003). Informativeness of genetic markers for inference of ancestry. *Am. J. Hum. Genet.*, **73**(6), pp. 1402–1422.

56. Smith, M. W., Lautenberger, J. A., Shin, H. D., Chretien, J. P., Shrestha, S., Gilbert, D. A., and O'Brien, S. J. (2001). Markers for mapping by admixture linkage disequilibrium in African American and Hispanic populations. *Am. J. Hum. Genet.*, **69**(5), pp. 1080–1094.

57. Anderson, E. C. and Thompson, E. A. (2002). A model-based method for identifying species hybrids using multilocus genetic data. *Genetics*, **160**(3), pp. 1217–1229.

58. Zhao, J., Boerwinkle, E., and Xiong, M. (2005). An entropy-based statistic for genomewide association studies. *Am. J. Hum. Genet.*, **77**(1), pp. 27–40.

59. Zhao, J., Boerwinkle, E., and Xiong, M. (2007). An entropy-based genome-wide transmission/disequilibrium test. *Hum. Genet.*, **121**(3–4), pp. 357–367.

60. Li, Y., Xiang, Y., Deng, H., and Sun, Z. (2007). An entropy-based index for fine-scale mapping of disease genes. *J. Genet. Genomics*, **34**(7), pp. 661–668.

61. Dong, C., Chu, X., Wang, Y., Jin, L., Shi, T., Huang, W., and Li, Y. (2008). Exploration of gene-gene interaction effects using entropy-based methods. *Eur. J. Hum. Genet.*, **16**(2), pp. 229-235.

62. Andrew, A. S., Nelson, H. H., Kelsey, K. T., Moore, J. H., Meng, A. C., Casella, D. P., Tosteson, T. D., Schned, A. R., and Karagas, M. R. (2006). Concordance of multiple analytical approaches demonstrates a complex relationship between DNA repair gene SNPs, smoking and bladder cancer susceptibility. *Carcinogenesis*, **27**(5), pp. 1030–1037.

63. Calle, M. L., Urrea, V., Vellalta, G., Malats, N., and Steen, K. V. (2008). Improving strategies for detecting genetic patterns of disease susceptibility in association studies. *Stat. Med.*, **27**(30), pp. 6532–6546.

64. Shervais, S., Kramer, P. L., Westaway, S. K., Cox, N. J., Zwick, M. (2010). Reconstructability analysis as a tool for identifying gene-gene interactions in studies of human diseases. *Stat. Appl. Genet. Mol. Biol.*, **9**(1), Article 18.

65. Zwick, M. (2004). An overview of reconstructability analysis. *Kybernetes*, **33**(5/6), pp. 877–905.

66. Cox, N. J., Frigge, M., Nicolae, D. L., Concannon, P., Hanis, C. L., Bell, G. I., and Kong, A. (1999). Loci on chromosomes 2 (NIDDM1) and 15

interact to increase susceptibility to diabetes in Mexican Americans. *Nat. Genet.*, **21**(2), pp. 213–215.

67. Horikawa, Y., Oda, N., Cox, N. J., Li, X., Orho-Melander, M., Hara, M., Hinokio, Y., Lindner, T.H., Mashima, H., Schwarz, P. E., del Bosque-Plata, L., Horikawa, Y., Oda, Y., Yoshiuchi, I., Colilla, S., Polonsky, K. S., Wei, S., Concannon, P., Iwasaki, N., Schulze, J., Baier, L. J., Bogardus, C., Groop, L., Boerwinkle, E., Hanis, C. L., and Bell, G. I. (2000). Genetic variation in the gene encoding calpain-10 is associated with type 2 diabetes mellitus. *Nat. Genet.*, **26**(2), pp. 163–175.

68. Tsalenko, A., Ben-Dor, A., Cox, N., and Yakhini, Z. (2003). Methods for analysis and visualization of SNP genotype data for complex diseases. *Pac. Symp. Biocomput.*, pp. 548–561.

69. Kasturi, J., Acharya, R., and Ramanathan, M. (2003). An information theoretic approach for analyzing temporal patterns of gene expression. *Bioinformatics*, **19**(4), pp. 449–458.

70. Hernandez-Lemus, E., Velázquez-Fernández, D., Estrada-Gil, J. K., Silva-Zolezzi, I., Herrera-Hernández, M. F., and Jiménez-Sánchez, G. (2009). Information theoretical methods to deconvolute genetic regulatory networks applied to thyroid neoplasms. *Phy. A: Stat. Mech. Appl.*, **388**(24), pp. 5057–5069.

71. Aldred, M. A., Huang, Y., Liyanarachchi, S., Pellegata, N. S., Gimm, O., Jhiang, S., Davuluri, R. V., de la Chapelle, A., and Eng, C. (2004). Papillary and follicular thyroid carcinomas show distinctly different microarray expression profiles and can be distinguished by a minimum of five genes. *J. Clin. Oncol.*, **22**(17), pp. 3531–3539.

72. Jarzab, B., Wiench, M., Fujarewicz, K., Simek, K., Jarzab, M., Oczko-Wojciechowska, M., Wloch, J., Czarniecka, A., Chmielik, E., Lange, D., Pawlaczek, A., Szpak, S., Gubala, E., and Swierniak, A. (2005). Gene expression profile of papillary thyroid cancer: sources of variability and diagnostic implications. *Cancer Res.*, **65**(4), pp. 1587–1597.

73. McGill, W. (1954). Multivariate information transmission. *Psychometrika*, **19**, pp. 97–116.

74. Matsuda, H. (2000). Physical nature of higher-order mutual information: intrinsic correlations and frustration. *Phys. Rev. E*, **62**(3), pp. 3096–3102.

75. Bell, A. (2003). The co-information lattice. In *The Fifth International Workshop on Independent Component Analysis and Blind Signal Separation (ICA 2003)*.

76. Leydesdorff, L. (2009). Interaction information: linear and nonlinear interpretations. *Int. J. Gen. Syst.*, **38**, pp. 681–685.

77. Matsuda, H. (2000). Physical nature of higher-order mutual information: intrinsic correlations and frustration. *Phys. Rev. E Stat. Phys. Plasmas Fluids Relat. Interdiscip. Top.*, **62**(3 Pt A), pp. 3096–3102.

78. Kirkwood, J. G. and Boggs, E. M. (1942). The radial distribution function in liqiuids. *J. Chem. Phys.*, **10**(6), pp. 394–402.

79. Krippendorff, K. (2009). Information of interactions in complex systems. *Int. J. Gen. Syst.*, **38**(6), pp. 669–680.

80. Leydesdorff, L. (2010). Redundance in systems which entertain a model of themselves: interaction information and the self-organization of anticipation. *Entropy*, **12**(1), pp. 63–79.

81. Zwick, M. (2011). Reconstructability analysis of epistasis. *Ann. Hum. Genet.*, **75**(1), pp. 157–171.

82. Clayton, D. G. (2009). Prediction and interaction in complex disease genetics: experience in type 1 diabetes. *PLOS Genet.*, **5**(7), p. e1000540.

83. Chanda, P., Zhang, A., and Ramanathan, M. (2011). Modeling of environmental and genetic interactions with AMBROSIA, an information-theoretic model synthesis method. *Hered. (Edinb.)*, **107**(4), pp. 320–327.

Chapter 3

Approaches for Gene–Environment Interaction Analysis: Practice of Regional Epidemiological Study

Mio Nakazato and Takahiro Maeda

Nagasaki University Graduate School of Biomedical Sciences,
205 Yoshikugi, Goto, Nagasaki 853-8691, Japan
mio@msb.biglobe.ne.jp, tmaeda@nagasaki-u.ac.jp

Since 2004 we have been conducting a regional epidemiological study on a remote Japanese island to clarify associations between genetic polymorphisms and lifestyle-related diseases. To facilitate this study, we established a research base with resident researchers in this study field, cooperated with the local government, and tried to build a relationship of mutual trust with the inhabitants. In this report, we describe the logistics of building a regional epidemiological study field, providing statistical analysis in the text. Furthermore, as the study advanced, researchers in other departments requested collaboration and wrote papers using our data and samples.

Our results suggested that remote places such as islands are appropriate for regional epidemiological research because the area

Gene–Environment Interaction Analysis: Methods in Bioinformatics and Computational Biology
Edited by Sumiko Anno

is limited and residents move infrequently. Furthermore, useful sample sizes and clear data in such studies will positively affect young doctors and researchers, who may advance their research careers in remote places. This might also contribute to better community medicine in remote places and islands.

3.1 Introduction

To smoothly and attentively promote a regional epidemiological study, cooperation between research institutions, such as universities, and communities is important, and it is desirable to establish a research base with resident research staff in the community. The Nagasaki University Graduate School of Biomedical Sciences opened the Department of Island and Community Medicine in May 2004 and simultaneously established the Island Medical Research Institute as an activity base of the department in an isolated island (Goto City) in the Nagasaki prefecture. This department was founded in Nagasaki University by donation from the local government, aiming to improve remote island medical care in the Nagasaki prefecture comprised of many islands, including the largest number of inhabited islands in Japan. The objectives of the department are community health care education, studies on regional epidemiology, sharing of community health care information, and medical care support for islands and remote rural areas. The staff of our group has been stationed in the institute since the opening, in 2004, and started and promoted regional epidemiological studies through coordination with related institutions in the community in cooperation with experts of epidemiological research. If young researchers starting studies refer to our methods, activate regional epidemiological studies, and contribute to improvement of community health care, nothing could give us more pleasure.

3.2 Community Investigation and Setting of the Research Objective

We considered that it is desirable to firstly know the regions in which an epidemiological study is to be performed, including the ethnic, cultural, and historical aspects, in addition to the medical aspect, through which the characteristics and problems of the region are clarified. We named this procedure *community diagnosis*. There is no specific mode or text for community diagnosis. Thus, we analyzed various survey reports published by the government, such as surveys of population and age composition, causes of death and requirement of long-term nursing, and industrial composition.

According to the 2009 survey document published by the hydrographic and oceanographic department, Japan is comprised of 6852 islands inhabited by about 1.2 hundred million people. As of 2010, the mean life expectancies of Japanese women and men are 86.30 and 79.55 years, respectively [1], being the world's longevity country. In the population composition in 2010, the percentage of the young population below 15 years was 13.2%, that of the working-age population between 15 and 64 years was 63.8%, and that of the elderly population aged 65 years or older was 23.0%, showing a higher rate of aging than in other countries. In the estimation for the future, the percentage of the elderly population is predicted to rise to 31.6% in 2030 and 38.8% in 2050, showing a high rate of aging (Table 3.1) [12].

Table 3.1 Annual changes in the percentage (%) of the elderly population in Japan and other countries

	Japan	UK	France	German	USA	Canada
1970	7.07	13.03	12.86	13.69	9.84	7.90
1990	12.08	15.70	14.03	14.89	12.49	11.27
2010	23.02	16.59	16.79	20.38	13.06	14.11
2030	31.60	21.11	23.10	28.03	19.91	22.95
2050	38.81	23.62	24.93	30.86	21.22	24.93

The subject field of our regional epidemiological study, Goto City in the Nagasaki prefecture, is located about 100 km from the mainland, and it consists of 11 isolated inhabited islands (Fig. 3.1). About 92,000 people were living on the islands in about 1960, when the population was the largest, but the population decreased to about 41,000 by 2010, and it is expected to decrease to about 28,000 in 2030. The percentage of the elderly population in 2010 was reported to be 33.4%, and this is similar to the predicted mean percentage for the nation after 20 years. Thus, the demographics of Goto City go ahead of nationwide aging. We considered that by selecting this region for the research field, we can investigate health problems of an aging society that Japan will face in the future.

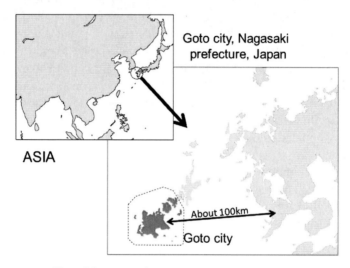

Figure 3.1   Map of the research areas.

As aging progresses and the elderly population increases, deaths from malignant neoplasms and lifestyle-related diseases, such as heart disease and cerebrovascular disorder, increase. According to the demographic statistics in 2010, 1,197,012 people died in Japan and the causes of death were malignant neoplasms, heart disease, cerebrovascular disorder, and pneumonia in 29.5%, 15.8%, 10.3%, and 9.9% of the cases, respectively.

In 2000, the World Health Organization (WHO) proposed an index, healthy life expectancy, representing the number of years that a person is expected to independently live in full health requiring no nursing care. The healthy life expectancies of Japanese women and men in 2010 were 73.6 and 70.4 years, respectively [2]. The duration of the period requiring nursing care, calculated from the difference from the mean life expectancy, is 12.8 years in women and 9.2 years in men. Regarding the causes of diseases requiring nursing care [3], cerebrovascular disorder is the most frequent cause (21.5%), followed by dementia (15.3%) and weakness due to aging (13.7%).

On the basis of the above, we considered that factors of cerebrovascular disorder and dementia leading to requirement of nursing care can be clarified by investigating lifestyle-related diseases, particularly arteriosclerosis, in aging-advanced regions, such as isolated islands, which is beneficial and significant for society. Concretely, we aimed at the establishment of a simple, low-invasive clinical evaluation method of arteriosclerosis applicable for screening, identification of risk factors (such as genes and lifestyles), extraction of the high-risk group, and establishment of an effective preventive method.

3.2.1 Study Background and Purpose: Homocysteine and Arteriosclerosis

In 1969, McCully [4–7] reported that hyperhomocysteinemia caused arteriosclerosis, as autopsies revealed many arteriosclerotic plaques in young patients with homocysteinuria, a congenital disease with a markedly high blood concentration of homocysteine. This substance is produced when methionine is metabolized to cysteine (Fig. 3.2).

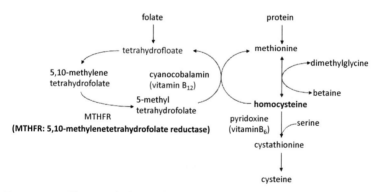

Figure 3.2   The metabolism of homocysteine and methylenetetrahydrofolate reductase.

In a basic study, Welch and Loscalzo (1997) [8] confirmed that homocysteine affected the vascular endothelium, enhancing platelets' coagulation capacity. Arteriosclerosis may be promoted through this mechanism. Thereafter, it was hypothesized that mild hyperhomocysteinemia in the absence of homocysteinuria might be a risk factor for arteriosclerosis. Many epidemiological studies were conducted. On the basis of the results, a study indicated the association between homocysteine and stenosis of the carotid artery, and another study reported that the blood concentration of homocysteine was significantly higher in patients with ischemic heart disease.

In addition, polymorphism of methylenetetrahydrofolate reductase (*MTHFR*), which is involved in homocysteine metabolism, was reported. Polymorphism refers to a gene mutation with an incidence of 1% or more. Gene mutations involving single-base replacement are defined as single-nucleotide polymorphisms (SNPs). In 1995, Frost et al. [9] reported that homotypic TT mutations involving mutation of the *MTHFR* gene 677 from C to T (*MTHFR*677C>T SNPs) reduced the enzymatic activity of *MTHFR* and increased the thermolability (Fig. 3.3), significantly elevating the blood concentration of homocysteine. Furthermore, de Bree et al. [10, 11] indicated that the consumption of a sufficient volume of folic acid decreased the blood concentration of homocysteine, even in patients with TT mutations in whom there was an increase in the blood concentration of homocysteine, suggesting that the genetic assessment of a high-risk group is possible and that folic acid prevents arteriosclerosis.

MTHFR Gene C677T polymorphism (rs1801133)

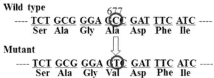

CC or CT : normal enzyme activity

TT   : reduced enzyme activity

Figure 3.3   *MTHFR* gene C677T polymorphism, the C-to-T substitution at nucleotide 677 converts an alanine to a valine residue at amino acid position222. The homozygous from of TT associated with reduced enzyme activity.

We established the following five study purposes:

- To analyze the association between homocysteine and parameters of arteriosclerosis
- To investigate the incidence of C677T polymorphism of *MTHFR* genes in Japanese
- To analyze the association between *MTHFR* polymorphism and parameters of arteriosclerosis

- To clarify factors influencing the blood concentration of homocysteine
- To examine regional differences in the blood concentration of homocysteine

Furthermore, we conducted a survey in three areas—small islands (a population of less than 500 residents), a large island (a population of approximately 40,000 residents), and the mainland (a population of 53,000 residents [13])—among which life environments may differ, utilizing the features of our field.

3.3   Study Design and Methods

After setting a study objective, it is necessary to investigate a study design and methods to actually perform the study. Close investigation of various items is necessary, such as what and how data or samples should be collected and how samples are to be processed at the time of collection and stored. Items considered important by us are shown below:

- Survey items
- Sample size
- Sample collection and storage
- Measurement and analysis of samples
- Data management
- Ethics committee
- Preparation of the field for regional epidemiology
- Investigation of the limitations and problems of the survey

3.3.1   Survey Items

The accuracy level and type of data may markedly influence factors such as cost, time, manpower, and place. In our survey, the plasma homocysteine concentration, evaluation of arteriosclerosis, and DNA were essential items to investigate the association of homocysteine with arteriosclerosis and genetic polymorphism. In addition, the severity of obesity, blood pressure, habitual cigarette smoking and alcohol drinking, past medical history of treatment, serum lipids, renal function, uric acid, and blood glucose, which are known to be associated with arteriosclerosis, were also surveyed within a possible range.

3.3.2   Sample Size

It becomes more likely to obtain a statistically significant finding as the sample size increases. Basically, it may be desirable to determine the number of necessary samples by performing a trial survey and on the basis of the statistical method expected to be used and the power. However, samples are collected within a range allowed by the budget, manpower, and time in many surveys.

3.3.3   Sample Collection and Storage

After the survey items and sample size are decided, a concrete sample collection method is investigated. It is necessary to perform close simulation of each survey item, starting from the participant recruitment method.

In our regional research survey, data and samples were collected at special health checkups of residents. These health checkups are operated by local governments for people insured by the national health insurance and the elderly aged over 75 years in Japan. Examinees have the right to undergo a mass checkup performed by an examiner group visiting each community or an individual checkup at a medical institution contracted for health checkups once a year. We visited mass checkups with the examiner group, registered examinees who gave consent as study participants, and collected data and samples, for which we made close arrangements with each organization from the preparatory step in the previous year and decided on the procedure by asking the public health department of the local government administrating health checkups for cooperation, explaining the study outline to the medical association; asking the contractor for mass checkups for cooperation; inspecting the mass checkup sites; selecting measurement devices; and preparing a combined flowchart of health checkup and survey.

3.3.3.1   Explanation of the community research survey to each organization and request for cooperation

To the public health department of the local government, we explained not only the academic significance of this regional epidemiological survey but also the merits for the regional public health department and community residents participating the study, and we could obtain consent for cooperation. Concretely, the following merits were presented: (i) participants can undergo measurement of indices of arteriosclerosis free of charge, (ii) data collected for the survey will be partially returned to the participants with an explanation, (iii) the health examination rate is expected to rise, and (iv) lectures on health and medical care using the survey data will be given as requested.

We selected physicians for health checkups in the communities to be surveyed, because mass health checkup contractors were struggling with a shortage of physicians for health examination. We also explained the outline of the study to the regional medical association prior to the study because participants in whom arteriosclerotic findings are detected on this health checkup may become anxious and consult with their attending physicians, which may confuse clinical sites unless the outline of the study is informed. Once a problem occurs in a regional survey, it may cause major damage and continuation of the survey may become difficult because cooperation cannot be obtained. Thus, these considerations are necessary. We also explained the outline in the early phase of the study prior to the survey to influential persons in the community, such as the chairperson of a club for the aged.

3.3.3.2 Methods and flow of the research survey

Resident health checkups were performed in the following order: reception; urine collection; measurement of the height, body weight, and abdominal circumference; interview; obtainment of consent; blood pressure measurement; blood sampling; physical examination; and optional cancer checkup (chest and upper gastrointestinal X-ray radiography). Since 50–150 residents undergo examination per half day in a mass health checkup, time-consuming measurement cannot be performed, and consideration of the installation area and invasiveness was also necessary. In addition to data collected by health checkups, we measured the body fat percentage, cardio-ankle vascular index (CAVI), and carotid intima-media thickness (CIMT) using carotid arterial echography. The body fat percentage could be measured without increasing the measurement time and installation area by changing height and weight scales to a device capable of simultaneously measuring the height, body weight, and body fat. The CAVI represents arterial stiffness from the heart through the ankle, similar to the pulse wave velocity (PWV), and it is less likely to be influenced by blood pressure during measurement compared to the PWV. Measurement was performed in a recumbent position. The CAVI can be measured within three minutes after attachment of manchettes to the four limbs, electrodes to the bilateral upper limbs, and a cardiac sound microphone to the chest. Since blood pressure of the four limbs can be measured at the same time, we measured CAVI instead of blood pressure. For CIMT measurement, the common carotid artery was imaged by ultrasonography using a 10 MHz linear probe, and the intima-media thickness was measured (Fig. 3.4). Using this, arteriosclerosis can be visually detected as the thickening and unevenness of the arterial wall in the image. We added the CIMT measurement to checkups. Since the measurement took only one to two minutes per person, it did not prolong the waiting time in the health checkups.

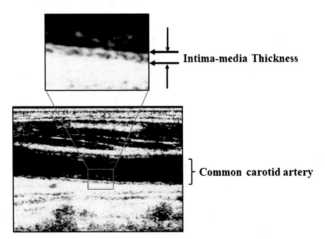

Figure 3.4   The intima-media thickness in the image of echographic examination.

3.3.3.3   Collection and storage of blood samples

Ethylenediaminetetraacetic acid (EDTA)-added blood (2 mL) for blood cell counting, plain blood (7 mL) for general biochemistry, and sodium fluoride–added blood (2 mL) for a blood glucose test are originally collected in resident health checkups. Blood samples additionally necessary for our study were 300 μL of plasma for homocysteine measurement and 1.5 mL of blood each for blood cell preparation for DNA extraction and analysis and plasma and serum stocks in preparation for later special tests. In preparation of plasma and blood cells, 5 ml of EDTA-added blood was additionally collected, and plasma for homocysteine measurement, blood cell components for DNA extraction, and plasma for stock were aliquoted. Regarding serum, residual serum samples after general biochemistry on health checkups were collected from the testing company as stocks. Each sample was contained in a numbered cryogenic or Eppendorf tube and stored at –30°C. Since a larger number and many types of storage tubes are used in a regional epidemiological study, the use of a label printer for numbering saves labor at a relatively low cost. It is also necessary to prepare a database of storage places easily accessible, as needed.

3.3.4 Measurement and Analysis of Samples

Special tests and gene analysis are performed after samples are accumulated to some extent in many surveys. Tests are performed by testing companies, or measurement and analysis are performed by us, depending on the budget, equipment, and manpower.

In our community survey, general blood test items of health checkups were measured by a clinical test company specified by the contract health checkup company. We planned to utilize these test data for our study after obtaining consent from the participants.

The plasma homocysteine concentration was measured in cryo-preserved plasma using high-performance liquid chromatography (HPLC) with fluorescence detection [19, 28]. This measurement was performed by a cooperative researcher good at analysis in the Department of Pharmacology. Assignment of sample collection, measurement, and analysis through a collaborative study may be essential to perform a regional epidemiological survey.

Regarding *MTHFR*677C>T SNPs, DNA was extracted from blood cell components (QuickGene-810, FUJIFIRM, Japan), and the SNPs were analyzed using a real-time polymerase chain reaction (PCR) (LightCycler 1.5, Roche Diagnostics, Switzerland). The current SNP analysis is very simple. After confirming the target SNPs in GenBank (http://www.ncbi.nlm.nih.gov/genbank/), 100 bases or the rs numbers before and after the target region are presented to a gene analysis company. They design primers and probes, calculate and report the T_m values (melting temperature), and synthesize these. SNPs can be analyzed by setting the temperature, time, and number of cycles of the PCR and the temperature and temperature change velocity of the melting curve in a LightCycler referring to the T_m value. Once conditions for analysis with no problem are established, stable results can be obtained by routine work, which may be entrusted to a part-time worker.

For *MTHFR*677C>T SNPs (rs1801133) analyzed by us, 5′-TGGCAGGTTACCCCAAG-3′ and 5′-TGATGCCCATGTCGGTGC-3′ were used as the forward and reverse primers, respectively, and 165-base DNA products before and after the target C677T SNPs were synthesized by 40 cycles of PCR. Two probes hybridized to the proximity of the products were prepared, and SNPs were analyzed using these. Probe 1 was labeled with a donor dye, fluorescein. Probe 2 contained the target SNP and was labeled with an acceptor dye, LC-Red. When these hybridize to regions close to each other, LC-Red emits fluorescence due to fluorescence resonance energy transfer (FRET), but when the temperature is elevated, the probe is released from the DNA product, causing no FRET, and thus, no fluorescence is emitted. In a melting curve prepared by plotting the temperature and fluorescence emission, fluorescence emission decreases at a lower temperature in the mutant than in the wild type because Probe 2 is readily released from the DNA product due to the mismatch of the mutation site in the hybridization, that is, the DNA product is the homotype mutant when fluorescence emission decreases at a low temperature in the melting curve, the wild homotype mutant when the emission decreases at a high temperature, and the heterotype mutant when a slope representing reduction of fluorescence emission is present at two points—low and high temperatures. LightCycler1.5 is capable of simultaneously analyzing 32 samples, but blank, wild homotype, heterotype, and mutant homotype control samples should be included as indices to judge the accuracy of analysis on the basis of the melting curves of the controls. When the analytical result of the control was inconsistent with the expected result, we discarded the SNPs' analytical result and repeated measurement. We occasionally obtained unexpected analytical results of the controls. The most frequent error was fluorescence emission by the blank control. Since the original DNA is 1.1 trillion times increased through 40 cycles of PCR, a slight contamination results in a false positive. To reduce contamination as much as possible, we prepared reagents and samples using a fume hood and frequently changed gloves, but fluorescence emission by the blank control still occasionally occurred. We changed pure water and reagents to new lots or reduced the number of PCR cycles to 30 in some cases to cope with this problem.

3.3.5   Data Management

Various data can be collected in a regional epidemiology survey, such as life and past medical histories, height, body weight, blood pressure, general blood and special test findings, and indices of genetic polymorphisms of participants. Data can be directly input in spreadsheet software, such as Microsoft Excel, when the number of data to be managed is up to several hundreds. When the number is greater, database construction is necessary. Since manual input of data requires labor and causes mistyping, data should be acquired as electronic data as much as possible, and a function to import it into a database is essential. Recent measurement devices mostly have a function to output results as electronic data, and testing companies report electronic data on request.

There is no need to point out that the database of the study results is most valuable. However, it is easily lost due to failure of recording media, such as a hard disk storing the data, and recording media may suddenly break down without a sign. Moreover, the database of a regional epidemiological study is a large collection of personal information, and its leakage develops into a major problem. Therefore, backup and high security are needed.

In our regional epidemiological survey, various data were collected from about 500–1000 participants per year. Assuming that input and management of data of about 100 items are necessary per person, the amount of data is enormous. Thus, we managed the data using a database from the beginning using the database software FileMaker (FileMaker, Inc., USA). FileMaker is capable of intuitively preparing a card-type database, and it is advantageous that simultaneous input and editing work through multiple PCs are possible by commoditizing the file. It also has functions to import from and export to CSV and Excel files, which are necessary for regional epidemiological survey databases. Using FileMaker, we constructed two databases. One was for information capable of identifying individuals, such as name, address, and phone number, for which a password was set so that only the personal information controller could access. The other was for all data concerning the study. These two databases were connected through a key ID so that only the personal information controller could combine these.

These two databases were recorded daily as backups at a specified time in network-attached storage (NAS), and access to this NAS was also controlled by setting a password. We requested the health checkup and general-testing companies to report the results as electronic data and imported the data into the database. Electronic CAVI data were acquired as follows: the CAVI device was connected to the network and data collected in exclusive software (VSS-10, Fukuda Denshi, Japan)-activated PCs were exported as CSV files. The CIMT was taken out as a JPEG file of echo images and subjected to measurement using the intima-media thickness (IMT) analysis software Intima Scope (Media Cross Inc., Japan). The results were exported as a CSV file, acquiring electronic data.

Documents collected at the survey sites, such as medical interview sheets, written consent, and the printout of CAVI results, were collectively filed and stored in a locked cabinet.

By compiling a database, data necessary for analysis can be immediately exported as a list in Excel and the file can be directly subjected to analysis and preparation of graphs using statistical analysis software, such as IBM SPSS.

3.3.6 Ethics Committee

When a regional epidemiological study comes in sight, it is necessary to submit specified documents, such as the study protocol, explanatory document, and written consent, to the ethics committee of one's own institution and receive approval. The study protocol has to conform to the Declaration of Helsinki stipulating the ethical principles for humans. The Declaration of Helsinki specifies that collaborators have to participate in a study of their own will after receiving sufficient explanation; their privacy has to be protected; collection, analysis, storage, and reuse of data used for the study have to be clearly specified; and subjects have the right to withdraw from the study any time, for which consent to participate in the study is obtained after detailed explanation of the study and the rights of subjects.

It may be desirable to perform an additional survey, depending on the study results and later information, but it is difficult to obtain written consent at each time in regional epidemiological studies. Thus, it is desirable to explain items on the assumption of an additional survey beforehand. To efficiently explain the study protocol to the ethics committee, a flowchart should be prepared that facilitates an understanding of the protocol and may clarify a problem (Fig. 3.5). It should also be taken into consideration that several months are required to obtain approval of the study protocol, including gene analysis, from the ethics committee in many cases.

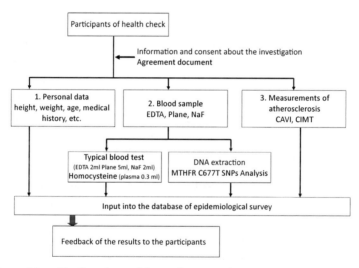

Figure 3.5   The flowchart of the study protocol.

3.3.7   Preparation of the Field of Regional Epidemiological Study

The field explored spending much effort is most valuable for the regional epidemiological study. To construct a favorable relationship with this field, giving sufficient explanation is important, as described above, but it is also important to send the results after the survey and hold a meeting to explain study results in the community. It is necessary to build mutual trust leading to the next survey through being involved in the field before, during, and after the survey.

We sent the arteriosclerosis test results and comments to individual participants, and the staff of our department attended and explained the results in a meeting for explanation of the health checkup results held by the local government. We also gave feedback of the study results to residents of the field by giving a lecture on health.

3.3.8   Study Limitations

The study protocol should be prepared so as to reduce biases as much as possible, but limitation occurs due to the budget and field elements.

In an ideal regional epidemiological study, it is desirable to survey all residents in the community or randomly extract subjects, but it is impossible for all target individuals to participate in the study. In our survey, the participants were those who gave consent out of persons who voluntarily underwent the health checkup. Therefore, they did not represent the general population of the community. This population may have been biased: persons having a strong interest in health, many elderly individuals with anxiety about health, and women with relatively much time during the day may have been included. Assuming that *MTHFR*677C>T SNPs strongly influence arteriosclerotic disease, persons with the disadvantageous polymorphism may have already died in the elderly population undergoing a health checkup, reducing the frequency of the disadvantageous polymorphism as a bias. In investigation of regional differences, the sample size markedly differs between communities with large and small populations. Although these limitations were considered, these may not have been problematic to investigate the association between homocysteine and the indices of arteriosclerosis in the population with high health consciousness and analyze differences between regions with different environments. On the basis of previous reports, we predicted that the influence of *MTHFR*677C>T SNPs on arteriosclerosis is not very strong.

3.4   Statistical Analysis

3.4.1   Preparation of Analytical Sheets

To perform statistical analysis, firstly, it is necessary to prepare analytical sheets. Our procedure is as follows: After export of data to be analyzed from the database, the data are opened in Excel and missing values and input errors are checked, for which AutoFilter is useful. By sorting in ascending and descending order each item, missing data, data in different formats, and apparent outliers can be found.

The exclusion criteria of samples are then determined referring to previous reports or on the basis of theoretical reasons. In our analysis, 2001 participants were included in analysis after excluding those with insufficient data collection. Since the objective was to investigate the association between subclinical arteriosclerosis and homocysteine, 106 subjects previously treated for arteriosclerotic diseases (ischemic heart disease, stroke, and arteriosclerosis obliterans) were excluded. In addition, 46 subjects with diabetes (previously treated or with a 6.5% or higher HbA1C) and 4 subjects with renal dysfunction (previously treated or with a 20 mg/L or higher serum creatinine) were also excluded because these were reported to strongly influence the progression of arteriosclerosis. Exclusion of subjects with hypertension and dyslipidemia was considered, but they were not excluded because the number of cases decreased to half when they were excluded. Finally, 1845 subjects (1299 females and 546 males) were included in the analysis. Of the 1845 subjects, 134 were residents of small islands with a population less than 500, 1426 were residents of a large island inhabited by about 43,000 people, and 285 were residents of a suburban city on the mainland with a population of 53,000. Since indices of arteriosclerosis could not be measured under the same condition in the residents of the mainland, they were excluded from analysis of indices of arteriosclerosis.

3.4.2   Close Investigation of Data

After data sheets are prepared, each type of measurement data is investigated before statistical analysis. Firstly, it is necessary to judge the type of measurement data (Table 3.2).

Table 3.2   Type of measurement data

1. Interval scale	Quantitative data with a clearly specified order and distance of the data	Temperature and test data
2. Ordinal scale	Only the order defined, such as large or small	Judgment of the effect, such as aggravation, ineffective, and markedly effective, and rank data, such as the first, second, and third places
3. Categorical scale (nominal scale)	Categorical data without definition of the order (large or small)	Data such as male/female and smoker/nonsmoker

For interval data, investigation by preparing a list of calculated means and standard deviations is insufficient. Data are characterized by the size, spread, and distribution, and these can be represented with the mean, median, standard deviation, maximum and minimum values, and 25% and 75% points. However, histograms and box-and-whisker plots visually representing the characteristics of measurement data are more useful.

A histogram is a graph plotting the frequency on the vertical axis and class on the horizontal axis, and it is useful to visually evaluate the distribution. The frequency distribution of homocysteine is shown in Fig. 3.6 as an example. In this histogram, the median tends to be smaller than the mean, and the mean, median, and 25% and 75% values are closer to the minimum than to the maximum value, showing an asymmetric distribution with a long tail on the higher-value side. The characteristics can be immediately identified referring to the histogram.

In a box-and-whisker plot, a box covering the 25% to 75% percentiles is drawn, the median is plotted on the box, and the maximum and minimum values after excluding outliers and extreme values are presented with bars. This is a superior plot so that the size, spread, and distribution can be visually judged using this graph alone. A box-and-whisker plot of homocysteine is shown in Fig. 3.7. The whisker is long upward, and outliers and extreme values are present on the larger-value side, showing a distribution with a long tail on the higher-value side.

Histogram

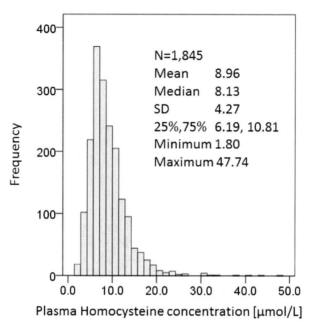

Figure 3.6   The histogram of plasma homocysteine concentration.

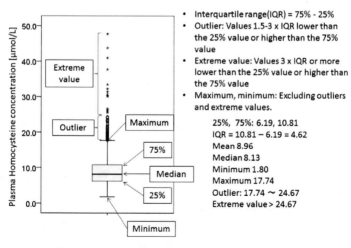

Figure 3.7   The box-and-whisker plot of plasma homocysteine concentration.

Histograms can be prepared using Excel but not in the default setting. In Excel 2010, click **File** → **Options** → **Add-ins** → **Manage: Excel Add-ins: Go**, and then select **Analysis Tool Pak** and click **OK** in the displayed window, and the condition for the use of histogram is set. Histograms can be prepared by selecting **Histogram** from **Data Analysis** in the **Data** tag. Box-and-whisker plots may be prepared using Excel, but it is necessary to calculate the maximum and minimum values, median, and 75% and 25% values of each measurement data type beforehand. After calculation, changing the format and drawing using **Cumulative Bar Chart** are necessary. To prepare histograms and box-and-whisker plots of many types of measurement data, it is very simple to use statistical analysis software. In SPSS, display the window by clicking **Analyze** → **Descriptive Statistics** → **Explore**, input the target measurement data into **Dependent List**, and select **Histogram** in **Plots**. Using this, the basic statistics, histogram, and box-and-whisker plot of many types of measurement data can be prepared.

According to the histogram, the distribution of homocysteine was not normal. An asymmetric distribution with a wide upward spread like this histogram is frequently noted in test values. There are the following approaches to cope with this condition:

- Use of a nonparametric test applicable for any type of distribution
- Transformation of the variable, such as logarithmic transformation, to make the distribution close to normal distribution

Analysts may select one of these. A histogram of homocysteine after logarithmic transformation is shown in Fig. 3.8 as an example. The mean and median are identical, and the distribution became symmetric. The data can be handled as those with normal distribution when the analyst visually judges the distribution as normal in the histogram, but when the sample number is small, it is difficult to evaluate the distribution on the basis of the histogram. Moreover, there may be interindividual variation in the judgment because subjectivity is present in visual evaluation. For this case, normality can be judged using the Kolmogorov–Smirnov and Shapiro–Wilk tests (Table 3.3). These two tests can also be applied using SPSS in the window displayed by clicking **Analyze** → **Descriptive Statistics** → **Explore**, and selecting **Normality plots with test** in **Plots**.

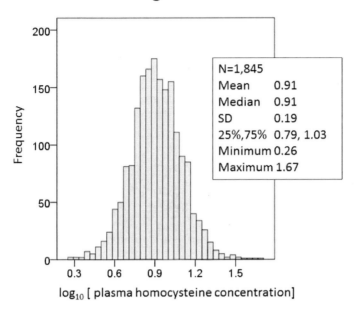

Figure 3.8 ── The histogram of plasma homocysteine concentration after logarithmic transformation.

Table 3.3 ── Normality test

	Kolmogorov–Smirnov Normality test			Shapiro–Wilk test		
	Statistics	Degrees of freedom	*p*	Statistics	Degrees of freedom	*p*
HCY	0.097	1845	<0.001	0.851	1845	<0.001
logHCY	0.021	1845	0.059	0.997	1845	0.001

HCY: plasma homocysteine concentration.

On the basis of these tests, homocysteine does not show normal distribution. Homocysteine after logarithmic transformation showed normal distribution in the histogram and on the Kolmogorov–Smirnov test but not on the Shapiro–Wilk test, showing inconsistency. It may be puzzling, but homocysteine after logarithmic transformation may be subjected to a parametric test as a normal distribution or nonparametric test as a nonnormal distribution, and it is better to quickly go on to the next analysis. Most of our measurement data did not show normal distribution on the Shapiro–Wilk test.

3.4.3 Selection of Statistical Analysis Methods

It is necessary to select statistical analysis methods on the basis of whether the measurement data are interval, ordinal, or categorical data and, when the data are interval data, whether the data have normality and homogeneity.

Representative statistical analysis methods considered useful for regional epidemiological studies are introduced below.

- Comparison of mean and frequency (Table 3.4)

Table 3.4   The methods at comparison of mean and frequency

Dependent variable	Interval scale		Ordinal scale	Categorical scale
	Normal and homogenous distribution	Non-normal or heterogeneous distribution		
	Parametric test	Nonparametric test		
Paired comparison of 2 group	paired t-test	Wilcoxon signed-rank test		sign test
Comparison of 2 independent groups	t-test	Mann-Whitney's U test		2 × 2 contingency table chi-square test Fisher's exact test
Paired comparison of 3 or more groups	two-way analysis of variance	Friedman's test		l x m contingency table chi-square test
Comparison of 3 independent groups	one-way analysis of variance	Kruskal-Wallis test		

- Association between variables
  - o Single regression analysis
  - o Multiple regression analysis
  - o Data grouped, such as division into quartiles, and compared
- When a confounding factor is present (multivariate analysis)
  - o When the target variable is quantitative (interval scale)
    - Analysis of covariance
    - General linear model
  - o When the target variable is binary
    - Logistic regression analysis

3.4.4   Practice of Analysis

One of the objectives of our survey was to investigate the association between the blood homocysteine level and indices of arteriosclerosis—CAVI and IMT—for which we firstly performed simple regression analysis of the relationships of the homocysteine level with the CIMT and CAVI (Fig. 3.9).

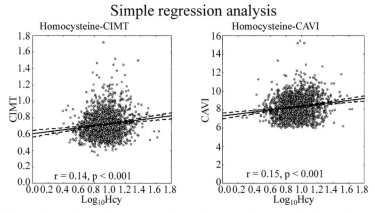

Figure 3.9   The scatter plots and simple regression analysis about the relationships of the homocysteine level with the CIMT and CAVI.

The homocysteine level was significantly correlated with the CIMT ($r = 0.14, p < 0.001$) and CAVI ($r = 0.015, p < 0.001$). Generally, the strength of correlation and the absolute value $|r|$ of correlation coefficient, r, are regarded as follows:

$0.7 \leq |r| \leq 1$ Strong correlation
$0.4 \leq |r| \leq 0.7$ Moderate correlation
$0.2 \leq |r| \leq 0.4$ Weak correlation
$|r| < 0.2$ No correlation

On simple regression analysis, the homocysteine level was statistically and significantly correlated with the CIMT and CAVI, but the correlation was very weak or absent.

However, the influence of a confounding factor is not considered in simple regression analysis, suggesting that association was not correctly judged because of the presence of other factors (sex, age, blood pressure, cholesterol, and HbA1C) associated with the indices of both arteriosclerosis and homocysteine. Thus, we performed multiple regression analysis regarding the CIMT and CAVI as dependent variables and measured data—age, sex, body mass index (BMI), systolic blood pressure (SBP), diastolic blood pressure (DBP), triglyceride (TG), high-density lipoprotein cholesterol (HDL-C), low-density lipoprotein cholesterol (LDL), HbA1C, creatinine, homocysteine, smoking status, and alcohol intake—as independent variables.

Multiple regression analysis explains and predicts a dependent variable with two or more independent variables. When appropriate independent variables are selected, a prediction equation with small errors can be prepared. Correlation between the independent variables is also taken into consideration in multiple regression analysis, and it attempts to investigate the influence of each independent variable on the dependent variable, excluding the influence of the correlation between the independent variables. Therefore, an independent variable with a significant partial regression coefficient against the dependent variable on multiple regression analysis may be significantly associated with the dependent variable even after eliminating the influence of the relationship between the independent variables. Using the standardized partial regression coefficient, the strength of the influence of independent variables on the dependent variable can be evaluated.

The following conditions are present for multiple regression analysis:

- Dependent variables form the interval data.

- For independent variables, interval data and dummy categorical data are applicable.
- When a strong correlation is present between independent variables, a regression coefficient contrary to the truth may be obtained.

Among the variables that we analyzed, sex, smoking status, and alcohol intake were transformed to dummy variables 0 and 1 because these variables are categorical data. Since the SBP and DBP were strongly correlated ($r = 0.78$, $p < 0.001$), the SBP was adopted as an independent variable because strong correlations with the CIMT and CAVI were noted on single correlation analysis, and the DBP was excluded.

For multiple regression analysis, it is necessary to prepare the optimum regression model by appropriately adding independent variables. Independent variables may be selected on the basis of previous knowledge, but selection employing the stepwise method or Akaike's information criterion (AIC) is also favorable.

Multiple regression analysis employing the stepwise method can be easily performed using SPSS: Click **Analyze** → **Regression** → **Linear**, specify dependent and independent variables, and execute by specifying **Stepwise** for **Method**. AIC is considered to be the general method to decide on a model by selecting variables, and it may be more preferred to the stepwise method. A smaller AIC number is favorable, and a model with the minimum AIC number is selected. In SPSS, the number can be calculated using syntax, but calculation has to be repeated until reaching the combination of the ideal model. Calculation is possible when the number of independent variables is small, but the number of combinations increases when there are many independent variables. Thus, calculation using SPSS is slightly complex compared to that using the stepwise method. For reference, the AIC is determined as follows in SPSS: Select **Analyze** → **Regression** → **Linear**, specify dependent and independent variables, and click **Paste** while keeping **Enter** for **Method**, and then Syntax Editor is activated and the following descriptions are displayed:

DATASET ACTIVATE DataSet1.
REGRESSION
    /MISSING LISTWISE

```
/STATISTICS COEFF OUTS R ANOVA
/CRITERIA=PIN(.05) POUT(.10)
/NOORIGIN
/DEPENDENT <independent variable>
/METHOD=ENTER <dependent variable1>
                        <dependent variable2>.....
```

Then, add **Selection** to the back of /STATISTICS COEFF OUTS R ANOVA to change it to /STATISTICS COEFF OUTS R ANOVA Selection and select **Run All** in Syntax Editor, and the calculation can be performed. Repeat the calculation, changing the combination of independent variables, to search for a combination leading to the minimum AIC number.

We investigated independent variables significantly associated with the CIMT or CAVI using the stepwise method (Table 3.5).

Table 3.5 Multiple regression analysis and stepwise procedure for the determinant of CIMT

	CIMT			CAVI		
	B	*β*	*p*	*B*	*β*	*p*
Age	0.006	0.431	<0.001	0.059	0.537	<0.001
Sex	–	–	–	0.163	0.061	0.007
BMI	–	–	–	−0.036	−0.096	<0.001
SBP	0.001	0.148	<0.001	0.007	0.115	<0.001
logTG	–	–	–	0.459	0.087	<0.001
HDL-C	–	–	–	–	–	
LDL-C	–	–	–	–	–	
HbA1C	–	–	–	0.135	0.045	0.031
Creatinine	0.012	0.234	<0.001	–	–	
$\log_{10}$ [homocysteine]	–	–	–	0.316	0.047	0.025
Smoking status	0.025	0.048	0.028	–	–	
Alcohol intake	–	–	–	0.193	0.068	0.002

CIMT: $r^2 = 0.29$, $r^2 = 0.38$.
B: partial regression coefficient.
β: standardized partial regression coefficient.
Age: year; sex: female 0 vs. male 1; BMI: kg/m^2; SBP: mmHg; logTG: log(g/L); HDL-C: g/L; LDL-C: g/L; HbA1C: %; $\log_{10}$ [homocysteine]: log(μg/L); CIMT: mm; CAVI: no unit; smoking status: noncurrent smoker 0 vs. current smoker 1; alcohol intake: none 0 vs. more than once a week 1.

Since the partial regression coefficient is the coefficient of each variable of the regression equation, the value markedly alters even in the same variable when the unit is changed. The strength of the influence of independent variables on the dependent variable can be presented employing the partial regression coefficient standardized to –1 to 1. The CIMT was significantly associated with age, SBP, creatinine, and smoking status, and CAVI was associated with age, sex, BMI, SBP, logTG, HbA1C, log homocysteine, smoking status, and alcohol intake. On the basis of the standardized partial regression coefficient, both CIMT and CAVI were strongly correlated with age, the CIMT was not associated with homocysteine, and CAVI was significantly associated with $\log_{10}$ [homocysteine]. This was a very weak association compared to that with age, and further investigation is necessary to judge whether this finding has actual meaning. Homocysteine was correlated with indices of arteriosclerosis in general healthy subjects in some recent reports, but no correlation was noted in others [14–17, 20–24].

Then, we investigated the results of *MTHFR*677C>T SNPs analysis. In this genetic polymorphism, *MTHFR* enzyme activity is normal in the CC and CT types, but it is reduced in the TT type, elevating the blood homocysteine level [18]. The frequency of *MTHFR*677C>T SNP was consistent with those in the Japanese in previous reports. We then performed multiple comparison of $\log_{10}$ [homocysteine] among the CC, CT, and TT types of *MTHFR*677C>T SNP. For comparison of three or more groups, you should not apply tests for two-group comparison, such as the *t*-test, to all comparative combinations, such as comparisons between the CT and CT types, between the CT and TT types, and between the CC and TT types, because of the following reason: When the mean is identical in three groups

$$\mu A = \mu B = \mu C$$

on a two-group comparison of three combinations, $\mu A = \mu B$, $\mu B = \mu C$, and $\mu A = \mu C$, when the acceptable level of probability of type I error, representing incidental misjudgment of identical means as different values, that is, the significance level, is set at 0.05, the probability of no occurrence of type I error is 0.95^3 in three tests of two-group comparison, resulting in the equation below:

Probability of occurrence of type I error in three tests of two-group comparison = $1 - (0.95)^3 = 0.143$.

The significant level increases, and a significant difference is likely to be detected despite no difference being present in truth. For comparison of the mean among three or more groups, the Bonferroni correction, one-way analysis of variance, or Kruskal–Wallis test is to be used.

The one-way analysis of variance and the Kruskal–Wallis test can be used to identify a significant difference among three or more groups, but between-group comparison, such as identification of the group with the highest value, cannot be performed. Basically, when between-group comparison is performed in tests of three or more groups, it may be better to apply multiple comparison (Table 3.6) from the beginning.

Table 3.6 Representative multiple comparison methods

Parametric test	Nonparametric test
Tukey–Kramer method	Steel–Dwass multiple comparison test
Dunnett's multiple comparison test	Shirley–Williams multiple comparison test
Scheffe test	
Bonferroni correction	

To perform multiple analysis using SPSS, select **Compare Means** → **One-Way ANOVA**, specify the target variable in the dependent list and categorical variable for the factor, and select a multiple analysis method in **Post Hoc** to execute the analysis.

Since the distribution of $\log_{10}$ [homocysteine] was judged as normal on the basis of the histogram and the Kolmogorov–Smirnov test, the Scheffe test was employed for multiple comparison (Table 3.7).

Table 3.7 Plasma homocysteine concentration among *MTHFR* 677C>T genotypes

MTHFR677C>T *n = 1845*	CC *n = 796*	CT *n = 859*	TT *n = 190*
log10 [homocysteine]	0.90 ± 0.18	0.91 ± 0.18	0.97 ± 0.21*

Values are expressed as the mean ± standard deviation.
*$p < 0.001$ between CC and TT; $p = 0.003$ between CT and TT.

On multiple comparison using the Scheffe test, no difference was noted between the CC and CT types but significant differences were observed between the CC and TT types and between the CT and TT types. These findings were consistent with those in previous reports. Thus, we designated the CC and CT types as a normal-enzyme-activity group and the TT type as a reduced-enzyme-activity group. The measurement data were compared between these two groups using Mann–Whitney test for interval data and the chi-square test for categorical data.

In SPSS, Mann–Whitney test can be performed by clicking **Analyze → Nonparametric Test → Legacy Dialogs → K Independent Samples**. The chi-square test can be performed by clicking **Analyze → Descriptive Statistics → Crosstabs** and selecting **chi-square** in **Statistics** (Table 3.8).

Table 3.8 The difference among *MTHFR* 677C>T SNPs

	*MTHFR*677C>T SNPs n = 1560		
	CC&CT n = 1399	TT n = 161	*p*
Age	63.4 ± 10.9	62.8 ± 11.7	0.92
Sex (female/male)	1000/399	105/56	0.10
BMI	23.1 ± 3.3	23.0 ± 3.2	0.58
Creatinine	7.2 ± 1.8	7.3 ± 1.8	0.33
Triglyceride	1.3 ± 0.8	1.3 ± 0.8	0.82
HDL-C	0.6 ± 0.2	0.6 ± 0.2	0.50
LDL-C	1.3 ± 0.3	1.2 ± 0.3	0.31
HbA1C	5.1 ± 0.4	5.1 ± 0.4	0.49
Homocysteine	9.2 ± 4.1	11.2 ± 5.6	<0.001
SBP	141.7 ± 20.1	139.2 ± 21.2	0.11
DBP	84.5 ± 11.1	83.4 ± 11.5	0.23
Smoking status (−/+)	1268/131	139/22	0.08
Alcohol intake (−/+)	1074/325	121/40	0.65
CIMT	0.7 ± 0.2	0.7 ± 0.2	0.82
CAVI	8.3 ± 1.2	8.3 ± 1.3	0.62

There was a significant difference in the homocysteine level between CC&CT and TT of *MTHFR677C>T* SNPs but no significant differences were noted in the age, sex, BMI, TG, HDL-C, LDL-C, HbA1C, SBP, DBP, CIMT, or CAVI, showing that *MTHFR677C>T* SNPs influence homocysteine but not the indices of arteriosclerosis. There was no variable in which a difference was noted between the CC&CT and TT types, other than homocysteine, suggesting that the presence of the CC&CT and TT types was compared in a homogenous population in our measurement data. On the basis of this, we concluded that there is no adjustment-requiring confound factor influencing the finding that there was no difference due to the *MTHFR677C>T* polymorphism in the CIMT or CAVI in our measurement data.

Finally, we investigated whether the homocysteine levels differ among the small islands, large island, and mainland, among which the living environment may be different. When $\log_{10}$ [homocysteine] was analyzed by multiple comparison using the Scheffe test (Fig. 3.10), the homocysteine level was significantly higher in the order of the small islands, large island, and mainland. However, in addition to the living environment, it is possible that the characteristics of residents are different, such as there are more elderly persons on the small islands, and it is necessary to investigate whether there is a confounding factor to be adjusted.

Plasma homocysteine concentration among small islands, large island and mainland.

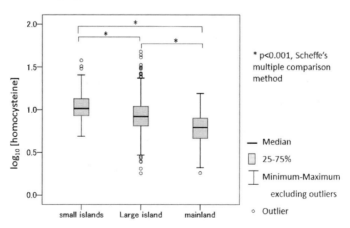

Figure 3.10 The box-wisker plots, the plasma homocysteine concentrations among residing areas.

The Kruskal–Wallis or chi-square test of each factor was performed in each region (Table 3.9).

Table 3.9 Characteristics of the study participants in each residing area

	Small islands (n = 134)	Large island (n = 1426)	Mainland (n = 285)
Age (year)[†]	67.7 ± 13.7	62.9 ± 10.6	61.0 ± 10.4
Male/female (%)[*]	40.3/59.7	28.1/71.9	31.9/68.1
BMI (kg/m²)[*]	23.8 ± 4.0	23.0 ± 3.2	23.3 ± 2.9
SBP (mmHg)[†]	148.8 ± 20.6	140.8 ± 20.8	128.7 ± 18.4
DBP (mmHg)[†]	85.9 ± 11.2	84.2 ± 11.1	77.5 ± 10.2
TG (g/L)[†]	1.27 ± 0.75	1.29 ± 0.80	1.09 ± 0.69
HDL-C (g/L)	0.57 ± 0.16	0.58 ± 0.15	0.58 ± 0.14
LDL-C (g/L)	1.24 ± 0.33	1.25 ± 0.33	1.22 ± 0.27
HbA1C (%)	5.1 ± 0.4	5.1 ± 0.4	5.1 ± 0.4
Creatinine (mg/L)[*]	6.8 ± 1.9	7.2 ± 1.8	7.0 ± 1.7
Homocysteine (μmol/L)[†]	11.5 ± 5.1	9.2 ± 4.2	6.5 ± 2.7
*MTHFR*677C>T (CC/CT/TT, %)	39.5/49.3/11.2	43.8/46.0/10.2	41.7/48.1/10.2
Smoking status (%)[†]	18.7	9.0	7.7
Alcohol intake (%)[†]	20.1	23.7	37.9

[*]$p < 0.05$; [†]$p < 0.001$ among three areas.

Since differences among the regions were noted in age, sex, BMI, SBP, DBP, TG, creatinine, homocysteine, smoking status, and alcohol intake, whether there is a regional difference in homocysteine should be concluded after adjustment for confounding factors. Thus, we identified factors influencing homocysteine by multiple regression analysis using the stepwise method, and factors showing differences among the three regions as well as influencing homocysteine were regarded as confounding factors for adjustment.

Table 3.10 Factors influencing logarithmic transformation of plasma homocysteine concentration

	Partial regression coefficient	Standardized partial regression coefficient	p
Age	0.002	0.086	$p < 0.001$
Sex	–	–	–
BMI	–	–	–
SBP	0.001	0.118	$p < 0.001$
TG	–	–	–
HDL-C	–0.115	–0.092	$p < 0.001$
LDL-C	–	–	–
HbA1c	–	–	–
Creatinine	0.026	0.249	$p < 0.001$
C677T/*MTHFR* (CC&CT vs. TT)	0.057	0.092	$p < 0.001$
Smoking status	0.085	0.133	$p < 0.001$
Alcohol intake	–0.023	–0.053	$p = 0.019$

$$r^2 = 0.16$$

Factors influencing $\log_{10}$ [homocysteine] were age, SBP, HDL-C, creatinine, *MTHFR*677C>T SNPs, smoking status, and alcohol intake. Of these, no differences were noted in HDL-C and *MTHFR*677C>T among the regions. Thus, age, SBP, creatinine, smoking status, and alcohol intake were regarded as confounding factors for adjustment.

We consulted a biostatistician on the method to analyze regional differences in homocysteine with adjustment for confounding factors and decided to employ analysis of covariance. Analysis of covariance can be performed using SPSS, but it is slightly complex. To perform it, use advanced statistical software with a biostatistician or refer to a manual. A similar analysis can be performed using the general linear model in SPSS. It may be better to consult a biostatistician on the results of analysis performed by yourself. We performed analysis of covariance with a biostatistician using SAS software version 9.13 (SAS Institute Inc., Cray, NC, USA).

Table 3.11 Logarithmic transformation of plasma homocysteine concentration among residing areas and adjusted for age, SBP, HDL-C, creatinine, smoking status, and alcohol intake

	Small islands (n = 134)	Large island (n = 1426)	Mainland (n = 285)
$\log_{10}$ [homocysteine]*	1.02 (0.01)	0.93 (0.01)	0.79 (0.01)

Values are adjusted means (standard errors) of logarithmic transformed homocysteine.

*p < 0.001 small islands vs. large island, small islands vs. mainland, and large island vs. mainland.

A significant difference was noted in the homocysteine levels despite no difference being present in the frequencies of *MTHFR*677C>T SNPs among the regions even after adjustment for the confounding factors. On multiple comparison, the homocysteine level was significantly higher in the order of the small islands, large island, and mainland. We reported these findings because a regional difference in the homocysteine levels within the same race has not previously been reported, although regional differences among different races have been reported. This regional difference in the homocysteine levels may be due to differences in dietary and exercise habits, which are environmental factors. Particularly, it is likely to be due to dietary habits because it has been reported that sufficient intake of folic acid reduces the blood homocysteine level in all genetic polymorphism types. On small islands with a population of less than 500 people, there may be only a distribution system and a traffic network, perishable food may not be sold in stores, and the meat intake may be low, with fish and shellfish being ingested mainly.

We went on one step further and performed analysis of covariance of *MTHFR*677C>T SNPs and regional differences.

The homocysteine level in the TT type was significantly higher in the small islands and the large island, and the difference tended to be larger in the small islands. There may be a lifestyle promoting *MTHFR*677C>T SNPs-associated hyperhomocysteinemia on the small islands.

Regional differences were noted in the blood homocysteine levels but not in the frequencies of *MTHFR*677C>T SNPs, suggesting that the cause was not the genetic factor, and it is likely to be the environmental factors. In addition, the presence of a lifestyle promoting the blood homocysteine–level elevation associated with the disadvantageous genetic polymorphism of *MTHFR677C>T* SNPs was suggested. However, homocysteine had no or only a small influence on the indices of arteriosclerosis, and *MTHFR677C>T* SNPs did not influence the indices of arteriosclerosis, although they influence the blood homocysteine level. These findings may be consistent with the result of a recent large-scale survey: folic acid supplement reduced the blood homocysteine level but did not reduce the incidence of arteriosclerotic disease [25].

Table 3.12 Logarithmic transformation of plasma homocysteine concentration among residing areas and *MTHFR677C>T* genotypes adjusted for age, SBP, HDL-Cm, creatinine, smoking status, and alcohol intake

C677T/MTHFR	Small islands (n = 134)	Large island (n = 1426)	Mainland (n = 285)
CC&CT (n = 1655)	1.00 (0.02)*†	0.92 (0.01)*†	0.80 (0.01)*
TT (n = 190)	1.18 (0.04)*†	0.98 (0.01)*†	0.77 (0.03)*

Values are adjusted means (standard errors) of logarithmic transformed homocysteine.
*p < 0.001 small islands vs. large island, small islands vs. mainland, and large island vs. mainland.
†p < 0.001 CC&CT vs. TT.

3.5 Later Surveys

Fields of regional epidemiological studies can be linked to various studies. Accumulated data may be analyzed from another viewpoint, previous studies may be further progressed, or a new study may be initiated on the basis of a new idea. If you have an attractive field and a database of regional epidemiological study, you will be requested by other researchers having ideas to perform a joint study.

3.5.1 Homocysteine and Folic Acid

As an association between homocysteine and folic acid consumption was suggested, the following study was conducted to clarify the association between homocysteine and folic acid intake using this regional epidemiological research field. Initially, a dietary survey was planned to investigate folic acid intake. However, a small number of preliminary surveys showed that a relatively strict weighing survey was necessary to evaluate the consumption of trace vitamins on the basis of the results of a dietary survey and that the sample size was restricted in a survey involving the small islands due to the small number of residents, raising important issues.

Subsequently, we planned to measure the blood level of folate. As folate in the blood, the serum or plasma level of folate, which reflects folic acid intake over a relatively short interval, or the erythrocyte level of folate, which reflects folic acid intake over a few months, is determined. Folate-measuring methods include the microbiological assay as a classical, standard method; the chemiluminescent immunoassay, which has recently been applied; radioimmunoassays; and high-performance liquid chromatography (HPLC). We performed a microbiological assay and a chemiluminescent immunoassay using the same samples. The two methods showed marked differences in measurements. For this analysis, two paired groups were tested using Wilcoxon's signed-rank test, as folate did not show a normal distribution. Subsequently, single regression analysis was conducted between the measurement methods. There were strong correlations in the serum or plasma and erythrocyte folate levels ($r = 0.85$, $p < 0.001$ [Fig. 3.11A] and $r = 0.87$, $p < 0.001$ [Fig. 3.11B], respectively). To classify the errors of differences related to measurements, we prepared Bland–Altman bias plots by plotting the mean of measurements per sample on the x axis and differences in measurements on the y axis. When there is a correlation between the x and y axes on the Bland–Altman bias plots (Fig. 3.11C), there may be a proportional bias of the systematic bias. When there is no correlation, with a mean Y value deviating from zero (Fig. 3.11D), the presence of a fixed bias of the systematic bias is suggested [27].

Furthermore, we conducted multiple regression analysis using the stepwise method to analyze factors influencing folic acid. There was an opposite correlation between the serum folate level and the BMI. However, there was no association between the erythrocyte folate level and the BMI. We published these results in a previous article [26].

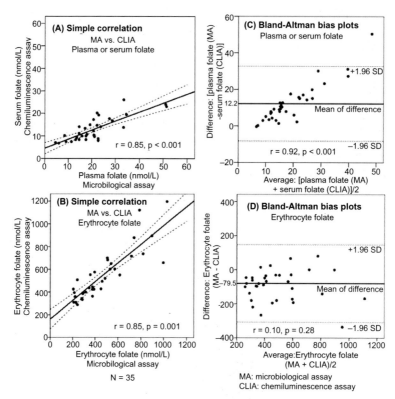

Figure 3.11   The scatter plots and Bland–Altman bias plots.

3.5.2   Adiponectin Polymorphism and Arteriosclerosis

Adiponectin is a cytokine secreted by fatty tissue. Obesity decreases the level of adiponectin. Adiponectin is associated with type II diabetes, coronary artery disease, and insulin resistance. The GG-type 276G>T SNP adiponectin gene shows higher insulin resistance and a decrease in the blood adiponectin level compared to the TT-type gene. Furthermore, adiponectin is classified into three types: high-, middle-, and low-molecular-weight adiponectin. In particular, high-molecular-weight adiponectin shows high activity, suggesting its association with insulin sensitivity. A collaborative investigator requested us to analyze 276G>T adiponectin gene polymorphism and high-molecular-weight adiponectin.

To investigate factors associated with high-molecular-weight adiponectin, multiple regression analysis was performed. However, in this study, analysis was conducted using high-molecular-weight adiponectin as a dependent variable and age/gender/target variables as independent variables. In other words, the influence of target variables on high-molecular-weight adiponectin was evaluated under age- and gender-adjusted conditions. The results of analysis showed that high-molecular-weight adiponectin was negatively correlated with the body weight, waist circumference, BMI, and triglyceride level. It was positively correlated with the HDL-C level. We compared high-molecular-weight adiponectin between the GG-type and GT&TT-type 276G>T SNP adiponectin genes. There was no significant difference. However, the results were adjusted with confounding factors such as age, gender, BMI, triglycerides, and HDL-C with respect to small and large islands and analyzed using covariance analysis. On the small islands, the GG-type high-molecular-weight adiponectin level was significantly lower. This is the first analysis of the 276G>T SNP adiponectin gene and high-molecular-weight adiponectin level; therefore, it was published as an article [29].

3.5.3   Leukocyte Count and Arteriosclerosis

A study reported the association between arteriosclerosis and high-sensitivity C-reactive protein (hsCRP), since hsCRP reflects inflammation, which plays an important role in the progression of arteriosclerosis. As a parameter of inflammation, the leukocyte count is also commonly used. A collaborative investigator wished to analyze the association between the leukocyte count and parameters of arteriosclerosis. The results of multivariate analysis by a biostatistician showed that there was a correlation between CAVI and the leukocyte count in males. However, there was no correlation in females. The CIMT was not correlated with the leukocyte count regardless of gender. This is the first analysis of the leukocyte count and parameters of arteriosclerosis; therefore, it was published as an article [30].

3.5.4   Current Study Purpose in Our Regional Epidemiological Research Field

Few studies have reported SNPs, which independently influence arteriosclerosis. On the other hand, many SNPs that are not associated with arteriosclerosis or may be weakly associated, such as *MTHFR*677C>T, have been reported. We are analyzing several SNPs that may be associated with arteriosclerosis in the future, scoring them, and collecting data on the basis of the hypothesis that the score may be associated with arteriosclerosis. Briefly, we are promoting research hypothesizing that the accumulation of several SNPs disadvantageous for arteriosclerosis may significantly increase the risk of arteriosclerosis although the influence of respective SNPs on arteriosclerosis is weak.

3.5.5   Other Studies

In addition to the above studies, the following themes were analyzed on the basis of the data obtained in this regional epidemiological research field and published as articles: association between cystatin C, which may be associated with chronic kidney disease, and parameters of arteriosclerosis; association between the leptin/ adiponectin ratio and parameters of arteriosclerosis; association between thyroid function and the CIMT; study regarding the usefulness of CAVI as an arteriosclerosis-screening tool; association between periodontal disease and parameters of arteriosclerosis; association between periodontal disease and HbA1C; and study regarding a homocysteine-measuring method. Thus, if a close, confident/cooperative relationship with the residents and administrative organizations is established through the arrangement of the regional epidemiological research field, data will be gradually collected, and investigators will gather to analyze the data. If young physicians/investigators working in the community can produce papers to earn a doctorate on the basis of regional epidemiological research, superior medical professionals may gather/grow in the community, contributing to advances in community health care. Furthermore, compact areas, such as remote places and islands on which residents' moves are not frequent, are appropriate for the regional epidemiological research field. This may contribute to resolving the lack of health care professionals in remote places and islands.

3.6 Conclusion

We have developed and arranged survey fields for lifestyle-related diseases. We felt that it was important to carefully obtain residents' and administrative organizations' cooperation, consult biostatisticians with respect to study protocols/analytical methods, and conduct investigational surveys in order to continuously investigate necessary items. We hope that our survey methods will help young investigators promote regional surveys.

References

1. Ministry of Health, Labour and Welfare, Japan. (2010). *The Summary of the 21st Life Table* (in Japanese).

2. Ministry of Health, Labour and Welfare, Japan. (2010). *2010 National Census* (in Japanese).

3. Ministry of Health, Labour and Welfare, Japan, *The Summary of 2010 Comprehensive Survey of Living Conditions* (in Japanese).

4. McCully, K. S. (1969). Vascular pathology of homocysteinemia: implications for the pathogenesis of arteriosclerosis. *Am. J. Pathol.*, **56**(1), pp. 111–128.

5. McCully, K. S. (1970). Importance of homocysteine-induced abnormalities of proteoglycan structure in arteriosclerosis. *Am. J. Pathol.*, **59**(1), pp. 181–194.

6. McCully, K. S., and Ragsdale, B. D. (1970). Production of arteriosclerosis by homocysteinemia. *Am. J. Pathol.*, **61**(1), pp. 1–11.

7. McCully, K. S. (1972). Homocysteinemia and arteriosclerosis. *Am. Heart. J.*, **83**(4), pp. 571–573.

8. Welch, G. N., and Loscalzo, J. (1997). Homocysteine and atherothrombosis. *N. Engl. J. Med.*, **338**, pp. 1042–1050.

9. Frosst, P., Blom, H. J., Milos, R., Goyette, P., Sheppard, C. A., Matthews, R. G., Boers, G. J., den Heijer, M., Kluijtmans, L. A., van den Heuvel, L. P., and Rozen, R. (1995). A candidate genetic risk factor for vascular disease: a common mutation in methylenetetrahydrofolate reductase. *Nat. Genet.*, **10**, pp. 111–113.

10. de Bree, A., Verschuren, W. M., Kromhout, D., Kluijtmans, L. A., and Blom, H. J. (2002). Homocysteine determinants and the evidence to what extent homocysteine determines the risk of coronary heart disease. *Pharmacol. Rev.*, **54**, pp. 599–618.

11. de Bree, A., Mennen, L. I., Zureik, M., Ducros, V., Guilland, J. C., Nicolas, J. P., Emery-Fillon, N., Blacher, J., Hercberg, S., and Galan, P. (2006). Homocysteine is not associated with arterial thickness and stiffness in healthy middle-aged French volunteers. *Int. J. Cardiol.*, **113**, pp. 332–340.

12. The World Bank, *World Development Indicators*, http://data. worldbank.org/data-catalog/world-development-indicators.

13. National Institute of Population and Social Security Research, Japan, *The Documents of 2010 Population Statistics in Japan* (in Japanese).

14. Perry, I. J., Refsum, H., Morris, R. W., Ebrahim, S. B., Ueland, P. M., and Shaper, A. G. (1995). Prospective study of serum total homocysteine concentration and risk of stroke in middle-aged British men. *Lancet*, **346**(8987), pp. 1395–1398.

15. Perry, I. J. (1996). Serum total homocysteine concentration and risk of stroke. *Lancet*, **348**(9040), p. 1526.

16. Chambers, J. C., Obeid, O. A., Refsum, H., Ueland, P., Hackett, D., Hooper, J., Turner, R. M., Thompson, S. G., and Kooner, J. S. (2000). Plasma homocysteine concentrations and risk of coronary heart disease in UK Indian Asian and European men. *Lancet*, **355**(9203), pp. 523–527.

17. Selhub, J., Jacques, P. F., Bostom, A. G., D'Agostino, R. B., Wilson, P. W., Belanger, A. J., O'Leary, D. H., Wolf, P. A., Schaefer, E. J., and Rosenberg, I. H. (1995). Association between plasma homocysteine concentration and extracranial carotid-artery stenosis. *N. Engl. J. Med.*, **322**, pp. 286–291.

18. Cronin, S., Furie, K. L., and Kelly, P. J. (2005). Dose-related association of *MTHFR* 677T allele with risk of ischemic stroke: evidence from a cumulative meta-analysis. *Stroke*, **36**, pp. 1581–1587.

19. Garcia, A. J., and Apitz-Castro, R. (2002). Plasma total homocysteine quantification: an improvement of the classical high-performance liquid chromatographic method with fluorescence detection of the thiol-SBD derivatives. *J. Chromatogr. B. Anal. Technol. Biomed. Life Sci.*, **779**, pp. 359–363.

20. Adachi, H., Hirai, Y., Fujiura, Y., Matsuoka, H., Satoh, A., and Imaizumi, T. (2002). Plasma homocysteine levels and atherosclerosis in Japan: epidemiological study by use of carotid ultrasonography. *Stroke*, **33**, pp. 2177–2181.

21. McQuillan, B. M., Beilby, J. P., Nidorf, M., Thompson, P. L., and Hung, J. (1999). Hyperhomocysteinemia but not the C677T mutation of methylenetetrahydrofolate reductase is an independent risk determinant of carotid wall thickening. The Perth Carotid Ultrasound Disease Assessment Study (CUDAS). *Circulation*, **99**, pp. 2383–2388.

22. Bots, M. L., Launer, L. J., Lindemans, J., Hofman, A., and Grobbee, D. E. (1997). Homocysteine, atherosclerosis and prevalent cardiovascular disease in the elderly: the Rotterdam Study. *J. Intern. Med.*, **242**, pp. 339–347.

23. Durga, J., Verhoef, P., Bots, M. L., and Schouten, E. (2004). Homocysteine and carotid intima-media thickness: a critical appraisal of the evidence. *Atherosclerosis*, **176**, pp. 1–19.

24. Mayer, O., Filipovský, J., Dolejsová, M., Cífková, R., Simon, J., and Bolek, L. (2006). Mild hyperhomocysteinaemia is associated with increased aortic stiffness in general population. *J. Hum. Hypertens.*, **20**, pp. 267–271.

25. Lonn, E., Yusuf, S., Arnold, M. J., Sheridan, P., Pogue, J., Micks, M., McQueen, M. J., Probstfield, J., Fodor, G., Held, C., and Genest, J. Jr. (2006). Homocysteine lowering with folic acid and B vitamins in vascular disease. *N. Engl. J. Med.*, **354**(15), pp. 1567–1577.

26. Nakazato, M., Maeda, T., Takamura, N., Wada, M., Yamasaki, H., Johnston, K. E., and Tamura, T. (2011). Relation of body mass index to blood folate and total homocysteine concentrations in Japanese adults. *Eur. J. Nutr.*, **50**(7), pp. 581–585.

27. Nakazato, M., Maeda, T., Emura, K., Maeda, M., Tamura, T. (2012). Blood folate concentrations analyzed by microbiological assay and chemiluminescent immunoassay methods. *J. Nutr. Sci. Vitaminol.*, **58**(1), pp. 59–62.

28. Ichinose, S., Nakamura, M., Maeda, M., Ikeda, R., Wada, M., Nakazato, M., Ohba, Y., Takamura, N., Maeda, T., Aoyagi, K., and Nakashima, K. (2009). A validated HPLC-fluorescence method with a semi-micro column for routine determination of homocysteine, cysteine and cysteamine, and the relation between the thiol derivatives in normal human plasma. *Biomed. Chromatogr.*, **23**, pp. 935–939.

29. Ishibashi, K., Takamura, N., Aoyagi, K., Yamasaki, H., Abiru, N., Nakazato, M., Kamihira, S., and Maeda, T. (2007). Multimers and adiponectin gene 276G>T polymorphism in the Japanese population residing in rural areas. *Clin. Chem. Lab. Med.*, **45**(11), pp. 1457–1463.

30. Sekitani, Y., Hayashida, N., Kadota, K., Yamasaki, H., Abiru, N., Nakazato, M., Maeda, T., Ozono, Y., and Takamura, N. (2010). White blood cell count and cardiovascular biomarkers of atherosclerosis. *Biomarkers*, **15**(5), pp. 454–460.

Suggested Readings

1. Ministry of Health, Labour and Welfare, Japan, *2010 Vital Statistics* (in Japanese).

2. Tango, T. (1998). *Statistical Sense* (Asakura, Japan) (in Japanese).

3. Nagata, Y. (2003). *How to Decide of the Sample Size* (Asakura, Japan) (in Japanese).

4. Nagata, Y. (1997). *The Basics of Statistical Multiple Comparison Method*, Michihiro Y. (ed.) (Scientist Press, Japan) (in Japanese).

5. Kenneth, J. R., Sander, G., and Timothy, L. L. (2007). *Modern Epidemiology*, 3rd Ed. (Lippincott Williams & Wilkins, USA).

6. Laurent, S., Boutouyrie, P., Asmar, R., Gautier, I., Laloux, B., Guize, L., Ducimetiere, P., and Benetos, A. (2001). Aortic stiffness is an independent predictor of all-cause and cardiovascular mortality in hypertensive patients. *Hypertension*, **37**, pp. 1236–1241.

7. Hoshino, A., Nakamura, T., Enomoto, S., Kawahito, H., Kurata, H., Nakahara, Y., and Ijichi, T. (2008). Prevalence of coronary artery disease in Japanese patients with cerebral infarction: impact of metabolic syndrome and intracranial large artery atherosclerosis. *Circ. J.*, **72**, pp. 404–408.

8. Nakamura, K., Tomaru, T., Yamamura, S., Miyashita, Y., Shirai, K., and Noike, H. (2008). Cardio-ankle vascular index is a candidate predictor of coronary atherosclerosis. *Circ. J.*, **72**, pp. 598–604.

9. Kadota, K., Takamura, N., Aoyagi, K., Yamasaki, H., Usa, T., Nakazato, M., Maeda, T., Wada, M., Nakashima, K., Abe, K., Takeshima, F., and Ozono, Y. (2008). Availability of cardio-ankle vascular index (CAVI) as a screening tool for atherosclerosis. *Circ. J.*, **72**, pp. 304–308.

10. Wall, R. T., Harlan, J. M., Harker, L. A., and Striker, G. E. (1980). Homocysteine-induced endothelial cell injury in vitro: a model for the study of vascular injury. *Thromb. Res.*, **18**, pp. 113–121.

11. Harker, L. A., Harlan, J. M., and Ross, R. (1983). Effect of sulfinpyrazone on homocysteine-induced endothelial injury and arteriosclerosis in baboons. *Circ. Res.*, **53**, pp. 731–739.

12. He, J. A., Hu, X. H., Fan, Y. Y., Yang, J., Zhang, Z. S., Liu, C. W., Yang, D. H., Zhang, J., Xin, S. J., Zhang, Q., and Duan, Z. Q. (2010). Hyperhomocysteinaemia, low folate concentrations and methylene tetrahydrofolate reductase C677T mutation in acute mesenteric venous thrombosis. *Eur. J. Vasc. Endovasc. Surg.*, **39**(4), pp. 508–513.

13. Outinen, P. A., Sood, S. K., Liaw, P. C., Sarge, K. D., Maeda, N., Hirsh, J., Ribau, J., Podor, T. J., Weitz, J. I., and Austin, R. C. (1998). Characterization of the stress-inducing effects of homocysteine. *Biochem. J.*, **332**, pp. 213–221.

14. Upchurch, G. R. Jr., Welch, G. N., Fabian, A. J., Freedman, J. E., Johnson, J. L., Keaney, J. F. Jr., and Loscalzo, J. (1997). Homocyst(e)ine decreases bioavailable nitric oxide by a mechanism involving glutathione peroxidase. *J. Biol. Chem.*, **272**, pp. 17012–17017.

15. Kobayashi, K., Akishita, M., Yu, W., Hashimoto, M., Ohni, M., and Toba, K. (2004). Interrelationship between non-invasive measurements of atherosclerosis: flow-mediated dilation of brachial artery, carotid intima-media thickness and pulse wave velocity. *Atherosclerosis*, **173**, pp. 13–18.

16. Potter, K., Hankey, G. J., Green, D. J., Eikelboom, J., Jamrozik, K., and Arnolda, L. F. (2008). The effect of long-term homocysteine-lowering on carotid intima-media thickness and flow-mediated vasodilation in stroke patients: a randomized controlled trial and meta-analysis. *BMC Cardiovasc. Disord.*, **20**, pp. 8–24.

17. Jacques, P. F., Bostom, A. G., Wilson, P. W., Rich, S., Rosenberg, I. H., and Selhub, J. (2001). Determinants of plasma total homocysteine concentration in the Framingham Offspring cohort. *Am. J. Clin. Nutr.*, **73**, pp. 613–621.

18. Vermeulen, E. G., Stehouwer, C. D., Twisk, J. W., van den Berg, M., de Jong, S. C., Mackaay, A. J., van Campen, C. M., Visser, F. C., Jakobs, C. A., Bulterjis, E. J., and Rauwerda, J. A. (2000). Effect of homocysteine-lowering treatment with folic acid plus vitamin B6 on progression of subclinical atherosclerosis: a randomised, placebo-controlled trial. *Lancet*, **355**(9203), pp. 517–522.

19. Nygård, O., Vollset, S. E., Refsum, H., Stensvold, I., Tverdal, A., Nordrehaug, J. E., Ueland, M., and Kvåle, G. (1995). Total plasma homocysteine and cardiovascular risk profile. The Hordaland Homocysteine Study. *JAMA*, **274**, pp. 1526–1533.

20. Dinavahi, R., Cossrow, N., Kushner, H., and Falkner, B. (2003). Plasma homocysteine concentration and blood pressure in young adult African Americans. *Am. J. Hypertens.*, **16**, pp. 767–770.

21. Fakhrzadeh, H., Ghotbi, S., Pourebrahim, R., Heshmat, R., Nouri, M., Shafaee, A., and Larijani, B. (2005). Plasma homocysteine concentration and blood pressure in healthy Iranian adults: the Tehran Homocysteine Survey (2003–2004). *J. Hum. Hypertens.*, **19**, pp. 869–876.

22. Hao, L., Ma, J., Zhu, J., Stampfer, M. J., Tian, Y., Willett, W. C., and Li, Z. (2007). High prevalence of hyperhomocysteinemia in Chinese adults is associated with low folate, vitamin B-12, and vitamin B-6 status. *J. Nutr.*, **137**, pp. 407–413.

23. Moore, S. E., Mansoor, M. A., Bates, C. J., and Prentice, A. M. (2006). Plasma homocysteine folate and vitamin B(12) compared between rural Gambian and UK adults. *Br. J. Nutr.,* **96**, pp. 508–515.

24. Casas, J. P., Bautista, L. E., Smeeth, L., Sharma, P., and Hingorani, A. D. (2005). Homocysteine and stroke: evidence on a causal link from mendelianrandomization. *Lancet,* **365**, pp. 194–196.

25. Nygård, O., Refsum, H., Ueland, P. M., Stensvold, I., Nordrehaug, J. E., Kvåle, G., and Vollset, S. E. (1997). Coffee consumption and plasma total homocysteine: the Hordaland Homocysteine Study. *Am. J. Clin. Nutr.,* **65**, pp. 136–143.

26. Gaume, V., Mougin, F., Figard, H., Simon-Rigaud, M. L., N'Guyen, U. N., Callier, J., Kantelip, J. P., and Berthelot, A. (2005). Physical training decreases total plasma homocysteine and cysteine in middle-aged subjects. *Ann. Nutr. Metab.*, **49**, pp. 125–131.

27. Herbert, V. (1990). Development of human folate deficiency. In *Folic Acid Metabolism in Health and Disease,* pp. 195–210, Picciano, M. F., Stokstad, E. L. R., and Gregory, J. F. III (eds.) (Wiley, New York).

28. Moor de Burgos, A., Wartanowicz, M., and Ziemlański, Ś. (1992). Blood vitamin and lipid levels in overweight and obese women. *Eur. J. Clin. Nutr.,* **46**, pp. 803–808.

29. Wahlin, Å., Bäckman, L., Hultdin, J., Adolfsson, R., and Nilsson, L. G. (2002). Reference values for serum levels of vitamin B12 and folic acid in a population-based sample of adults between 35 and 80 years of age. *Pub. Health Nutr.*, **5**, pp. 505–511.

30. Kant, A. K. (2003). Interaction of body mass index and attempt to lose weight in a national sample of US adults: association with reported food and nutrient intake, and biomarkers. *Eur. J. Clin. Nutr.*, **57**, pp. 249–259.

31. Mojtabai, R. (2004). Body mass index and serum folate in childbearing age women. *Eur. J. Epidemiol.*, **19**, pp. 1029–1036.

32. Lawrence, J. M., Watkins, M. L., Chiu, V., Erickson, J. D., and Petitti, D. B. (2006). Do racial and ethnic differences in serum folate values exist after food fortification with folic acid? *Am. J. Obstet. Gynecol.*, **194**, pp. 520–526.

33. Huemer, M., Vonblon, K., Födinger, M., Krumpholz, R., Hubmann, M., Ulmer H., and Simma, B. (2006). Total homocysteine, folate, and cobalamin, and their relation to genetic polymorphisms, lifestyle and body mass index in healthy children and adolescents. *Pediatr. Res.,* **60**, pp. 764–769.

34. Papandreou, D., Mavromichalis, I., Makedou, A., Rousso, I., and Arvanitidou, M. (2006). Reference range of total serum homocysteine level and dietary indexes in healthy Greek schoolchildren aged 6–15 years. *Br. J. Nutr.,* **96**, pp. 719–724.

35. Papandreou, D., Rousso, I., Makedou, A., Arvanitidou, M., and Mavromichalis, I. (2007). Association of blood pressure, obesity and serum homocysteine levels in healthy children. *Acta. Pædiat.,* **96**, 1819–1823.

36. Rhee, E. J., Hwang, S. T., Lee, W. Y., Yoon, J. H., Kim, B. J., Kim, B. S., Kang, J. H., Lee, M. H., Park, J. R., and Sung, K. C. (2007). Relationship between metabolic syndrome categorized by newly recommended by International Diabetes Federation criteria with plasma homocysteine concentrations. *Endocr. J.,* **54**, pp. 995–1002.

37. Aasheim, E. T., Hofsø, D., Hjelmesæth, J., Birkeland, K. I., and Bøhmer, T. (2008). Vitamin status in morbidly obese patients: a cross-sectional study. *Am. J. Clin. Nutr.,* **87**, pp. 362–369.

38. Elshorbagy, A. K., Nurk, E., Gjesdal, C. G., Tell, G. S., Ueland, P. M., Nygård, O., Tverdal, A., Vollset, S. E., and Refsum, H. (2008). Homocysteine, cysteine, and body composition in the Hordaland Homocysteine Study: does cysteine link amino acid and lipid metabolism? *Am. J. Clin. Nutr.,* **88**, pp. 738–746.

39. Mahabir, S., Ettinger, S., Johnson, L., Baer, D. J., Clevidence, B. A., Hartman, T. J., and Taylor, P. R. (2008). Measures of adiposity and body fat distribution in relation to serum folate levels in postmenopausal women in a feeding study. *Eur. J. Clin. Nutr.,* **62**, pp. 644–650.

40. Ortega, R. M., López-Sobaler, A. M., Andrés, P., Rodoríguez-Rodoríguez, E., Aparicio, A., and Perea, J. M. (2009). Folate status in young overweight and obese women: changes associated with weight reduction and increased folate intake. *J. Nutr. Sci. Vitaminol.,* **55**, pp. 149–155.

41. Tamura, T. (1990). Microbiological assay of folates. In *Folic Acid Metabolism in Health and Disease,* pp. 121–137, Picciano, M. F., Stokstad, E. L. R., and Gregory, J. F. III (eds.) (John Wiley, New York).

42. Tamura, T., Freeberg, L. E., and Cornwell, P. E. (1990). Inhibition by EDTA of growth of Lactobacillus casei in the folate microbiological

assay and its reversal by added manganese or iron. *Clin. Chem.,* **36**, p. 1993.

43. Tamura, T., Bergman, S. M., and Morgan, S. L. (1998). Homocysteine, B-vitamins, and vascular-access thrombosis in patients treated with hemodialysis. *Am. J. Kidney Dis.,* **32,** pp. 475–481.

44. Yang, Q., and Erickson, J. D. (2003). Influence of reporting error on the relation between blood folate concentrations and reported folic acid-containing dietary supplement use among reproductive-aged women in the United States. *Am. J. Clin. Nutr.,* **77**, pp. 196–203.

45. Winkels, R. M., Brouwer, I. A., Verhoef, P., van Oort F. V., Durga, J., and Katan, M. B. (2008). Gender and body size affect the response of erythrocyte folate to folic acid treatment. *J. Nutr.,* **138**, pp. 1456–1461.

46. van Driel, L. M., Eijkemans, M. J., de Jonge, R., de Vries, J. H., van Meurs, J. B., Steegers, E. A., and Steegers-Theunissen, R. P. (2009). Body mass index is an important determinant of methylation biomarkers in women of reproductive ages. *J. Nutr.,* **139**, pp. 2315–2321.

47. Rasmussen, S. A., Chu, S. Y., Kim, S. Y., Schmid, C. H., and Lau, J. (2008). Maternal obesity and risk of neural tube defects: a metaanalysis. *Am. J. Obstet. Gynecol.,* **198**, pp. 611–619.

48. Shaw, G. M., and Carmichael, S. L. (2008). Prepregnancy obesity and risk of elected birth defects in offspring. *Epidemiology,* **19**, pp. 616-2-620.

49. Tamura, T., and Picciano, M. F. (2006). Folate and human reproduction. *Am. J. Clin. Nutr.,* **83**, pp. 993–1016.

50. Tamura, T., Goldenberg, R. L., Johnston, K. E., and Chapman, V. R. (2004). Relationship between pre-pregnancy BMI and plasma zinc concentrations in early pregnancy. *Br. J. Nutr.,* **91**, pp. 773–777.

51. Gunter, E. W., Bowman, B. A., Caudill, S. P., Twite, D. B., Adams, M. J., and Sampson, E. J. (1996). Results of an international round robin for seru and whole-blood folate. *Clin. Chem.,* **42**, pp. 1689–1694.

52. Garbis, S. D., Melse-Boonstra, A., West, C. E., and van Breemen, R. B. (2001). Determination of folates in human plasma using hydrophilic interaction chromatography-tandem mass spectrometry. *Anal. Chem.,* **73**, pp. 5358–5364.

53. Pfeiffer, C. M., Fazili, Z., McCoy, L., Zhang, M., and Gunter, E. W. (2004). Determination of folate vitamers in human serum by stable-isotope-dilution tandem mass spectrometry and comparison with radioassay and microbiologic assay. *Clin. Chem.,* **50**, pp. 423–432.

54. Kok, R. M., Smith, D. E., Dainty, J. R., van Den Akker, J. T., Finglas, P. M., Smulders, Y. M., Jakobs, C., and De Meer, K. (2004). 5-Methyltetrahydrofolic acid and folic acid measured in plasma with liquid chromatography tandem mass spectrometry: applications to folate absorption and metabolism. *Anal. Biochem.,* **326**, pp. 129–138.

55. Shane, B., Tamura, T., and Stokstad, E. L. (1980). Folate assay: a comparison of radioassay and microbiological methods. *Clin. Chim. Acta.,* **100**, pp. 13–19.

56. Owen, W. E., and Roberts, W. L. (2003). Comparison of five automated serum and whole blood folate assays. *Am. J. Clin. Pathol.,* **120**, pp. 121–126.

Chapter 4

Use of Bioinformatics in Revealing the Identity of Nature's Products with Minimum Genetic Variation: The Sibling Species

K. Gajapathy,[a] A. Ramanan,[b] S. L. Goodacre,[c] R. Ramasamy,[d] and S. N. Surendran[a]

[a]Department of Zoology, Faculty of Science, University of Jaffna 40000, Sri Lanka
[b]Department of Computer Science, Faculty of Science, University of Jaffna 40000, Sri Lanka
[c]School of Life Sciences, University of Nottingham, Nottingham NG7 2RD, UK
[d]Department of Biomedical and Forensic Sciences, Faculty of Science and Technology, Anglia Ruskin University, Cambridge CB1 1PT, UK
gayan156@gmail.com, gajapathyk@jfn.ac.lk, a.ramanan@jfn.ac.lk, sara.goodacre@nottingham.ac.uk, ranjan.ramasamy@anglia.ac.uk, noble@jfn.ac.lk

In a world where increasingly powerful computers are becoming available and are being made use of in science, we reflect on their use in taxonomy. Taxonomy is where we witness a grand stage show run by the environment and how nature shapes diversity among living organisms. Taxonomy has experienced many problems throughout the history of mankind, including the cryptic species where the

Gene–Environment Interaction Analysis: Methods in Bioinformatics and Computational Biology
Edited by Sumiko Anno
Copyright © 2016 Pan Stanford Publishing Pte. Ltd.
ISBN 978-981-4669-63-4 (Hardcover), 978-981-4669-64-1 (eBook)
www.panstanford.com

interaction between the environment and genes is delicate and thus produces a variant with minimum genetic and/or morphological variations. The chapter aims to assess the use of the power of computers to help in the differentiation of sibling species, especially among insect vectors of human diseases, which is often important for developing effective vector control methods.

4.1   Cryptic Species: An Introduction

The interaction between the environment and genes is always a hot topic. The environment shapes the biology of an organism, with various degrees of success, as some organisms disappear without a trace, while others succeed and thrive for a long time. The variations created by such interactions don't always give birth to a new species; instead they, at times, create variants with minimum variations, either in the gene structure or in the morphology.

Cryptic species are defined as two or more species that have been erroneously classified as one species because they are hard to distinguish on the basis of morphology alone. However, molecular or behavioral variations are possible ways to differentiate them. Sibling species (cryptic sister species) are always cryptic species. They are also morphologically indistinguishable. But the only difference is that sibling species are sister clades in modern phylogenetic relationships based on the alignment of molecular data (DNA or amino acid sequences), where that relationship is not necessary in cryptic species [1]. The existence of sibling species was recorded in the pre-Linnaean era. William Derham was credited as the first person to recognize sibling species among the bird genus *Phylloscopus* in 1718 [2]. Sibling species often provide useful systems in which to study the diversification events that occur during the process of evolution, because in many cases, but not all, they have evolved quite recently [3]. Sibling species have, by definition, a close phylogenetic relationship, but they no longer freely exchange genes with one another. This allows us to explore the biological species concept, which emphasizes reproductive isolation as a central part of the speciation process.

Beyond this, the existence of sibling/cryptic species can affect conservation measures, as in the case of marine turtles [4], lemurs

[5], and amphibians [6]. Natural selection can play a significant role in driving the evolution of particular genotypes and phenotypes. "Ecological speciation" is the term often used to describe such a process when it leads to reproductive isolation between diverged populations and the generation of new species.

Some environments appear to result in morphological stasis (the cessation of morphological change) in the evolution/speciation among species. Examples of this are found in extreme environmental conditions, such as the Arctic [7]) and areas of deep oceans [8]. The phenomenon of cryptic species cannot be resolved in these areas by morphology alone. Conversely, the persistence of high levels of variability, such as variation in shell colors within and among invertebrates such as land snails in northern Europe, does not appear to be explained by environmental selection alone despite its known influence on heat tolerance [9].

The widespread occurrence of sibling and cryptic species in insect vectors is well established, especially in anopheline mosquitoes, where the number of sibling species with different vectorial capacity has been documented [10]. *Anopheles culicifacies* and *A. subpictus* are classic examples where one or two specific sibling species have vectorial capacity and are attributed as the major vector in some part of the world. But the most interesting thing is some other sibling species has vector potential in different regions. *Phlebotomus argentipes* morphospecies A has the vector potential over species B in Sri Lanka, whereas it is the other way round in India. All these differences occur within the minimum morphological variances they possess. The role of the environment in shaping such differential biology in very closely related species is very subtle and subject to much scrutiny.

Cryptic species are known to be common in insects that are vectors of human diseases. One complicating factor is that some sibling species occur in sympatry, where the environment is the same [11, 12]. They are good models for studying speciation because they have evolved relatively recently. Straightforward predictions about the environment changes that have governed those changes can also be made. The pattern of changes in either genotype and/or phenotype might be easy to understand in the otherwise normal speciation process.

The most common attribute among the sibling species among insect vectors is the subtle variation in morphology. The fact is proven critical in insect vectors such as mosquitoes and sand flies as the differential vector potential and biology, including insecticide resistance, are common among them. The very minute variations found among the insect vectors in our research have caused many problems, which will be discussed in the next sections.

4.2 Computer Power in Taxonomy of Sibling Species

Bioinformatics, the integration of biology and computer applications, is used in many ways in modern taxonomy, from processing images (morphology, karyotyping, and chromosome mapping) to a more complex approach of reconstructing phylogenetic relationships among organisms. Here we discuss potential areas in taxonomy where computational approaches can be useful.

4.2.1 DNA and Protein as Tools in Taxonomy: Alternatives or Auxiliary?

On a global scale, species diversity is huge, with some studies estimating that there are 10–15 million species [13, 14]. Discrimination of them up to the species level demands a large number of expert taxonomists. This has become a huge task, given the recent decline in the number of taxonomists [15]. Several limitations in taxonomy have been cited. The limitations include the phenotypic and genotypic variances in desired traits, the morphologically cryptic taxa that are often oversighted, taxonomic keys that are much more specific for a gender or life stages, and the requirement of a high level of expertise [15].

The suggestion of using "microgenomics" (a term used in Ref. [15]), or studying the small fractions of the genome as an alternative to morphology-based taxonomy, was later criticized by Will and Rubinoff [16]. Will and Rubinoff [16] strongly denied the limitations of morphology-based analysis in taxonomy. Phenotypic variations corresponding with environmental gradients and phenotypic plasticity are well-known phenomena, having been observed in many organisms that have been the basis of taxonomic research for more than two centuries, since the Linnaeus binomial nomenclature in the 1750s. It seems unlikely that organisms can survive changing biotic or abiotic environments without changes in genotype and phenotype.

Phenotypic changes can have both heritable and nonheritable (i.e., plastic) components. Nonheritable variations in traits found among a taxon during assessments of their morphology are particularly problematic in terms of placement of that particular taxon within an existing taxonomic framework. This has been the experience during our sand fly survey in Sri Lanka. Many sand fly samples were identified up to the genus level and then one or two morphological features were different from the closest higher taxon. For instance, we found a sand fly specimen (female) from Padavisiripura village (in eastern Sri Lanka) that possesses many characters attributable to a very rare subgenus *Demeillonius* Davidson. The spermatheca structure was different in our specimen. Similarly the *A. subpictus* species from Sri Lanka showed much diverged morphology in the palp segments and were often identified as new variants (See Table 4.1 and Fig. 4.1), but this was not supported by the DNA sequencing results [17].

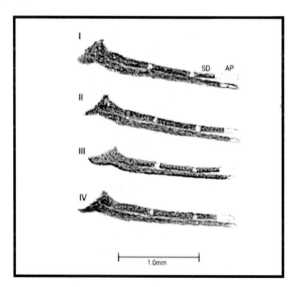

Figure 4.1 Variations in the ornamentation of palpi of *A. subpictus* species B. AP: apical pale band; SD: subapical dark band.

Table 4.1   Variations in the ornamentation of wings and palpi of the identified four populations of sibling species B of the *subpictus* complex

Type of population	No. of egg ridges	Variations in the ornamentation of	
		Wings	**Palpi**
I	17	Prehumeral dark spot complete, preapical dark spot on costa is long and larger than or equal to the length of the subcostal pale spot plus the preapical pale spot	The apical pale band larger than the subapical dark band
II	16–19	Prehumeral dark spot incomplete, preapical dark spot on costa short and shorter than the length of the subcostal pale spot plus the preapical pale spot	The apical pale band shorter than the subapical dark band
III	16–19	Prehumeral dark spot complete, preapical dark spot on costa short and shorter than the length of the subcostal pale spot the plus preapical pale spot	The apical pale band shorter than the subapical dark band
IV	16–18	Prehumeral dark spot complete, preapical dark spot on costa long and larger than or equal to the length of the subcostal pale spot plus the preapical pale spot	The apical pale band equal to the subapical dark band

Tautz et al. [18] were looking at DNA taxonomy as an auxiliary unit in mainstream taxonomy. They clearly indicated their respect toward morphology-based analysis, which is the core in any taxonomic analysis.

The genotypic variations suggested by Hebert et al. [15] were considered to be addressing the disparity between morphological and genetic data. One way or another, the evolution of a gene or a fragment of it need not be addressing the evolutionary pattern of the species. The pattern of evolution of a gene need not be simple and can be reticulated. Some genes may diverge rapidly, whereas others may be largely constrained by natural selection. This well-documented fact has been the focus of the argument by Will and Rubinoff [16]. They conclude it as a fundamental problem to phylogenetic but not pure taxonomy research. Other arguments have also been put forward to explain the disconnection between true species phylogenies and the phylogenetic trees produced for single genes. These include the persistence of ancestral polymorphisms and occasional hybridization events during or after the speciation process. Organisms with highly structured populations may be the most vulnerable to these types of phenomena.

Limitations in classical taxonomy due to cryptic taxa are considered as a special case in biology (e.g., Ref. [16]). There are ways to overcome the very little or no morphological differences among closely related cryptic species. Behavioral, physiological, morphometric, and biochemical approaches can be used as a basis for distinguishing among taxa. The most apt example can be using different life stages for assessing the taxonomy. *A. subpictus* mosquito adults carry very little variation in their morphology, while there is enough variation in their eggs to assess the differences (Fig. 4.2).

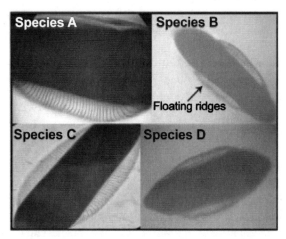

Figure 4.2 Egg ornamentation differences among the four sibling species in the *A. subpictus* species complex.

The other limitations include the availability of taxonomic keys for a single biological sex. This was addressed by Will and Rubinoff [16]. They argued that the lack of generalized keys is the same as the lack of data for all the genes for all the species or species groups. They also questioned the availability of gene sequence data of fossil and museum (we assume *rare* museum) specimens.

The fourth limitation suggested by Hebert et al. [15] is the need of expertise in analyzing the key and the misidentification due to the lack of it. Will and Rubinoff [16] accepted that morphological analysis is time consuming and the lack of expertise can cause a significant problem in the identification process. But they argue that misidentification is also possible in a highly simple (they have used the term "packaged kits" for extraction and polymerase chain reaction (PCR) in molecular biology) DNA-based technique. They point out the mismatches in the phylogeny trees illustrated by Hebert et al. [15].

Tautz et al. [18] were also concerned about the need for expertise in the morphology-based analysis. The lack of taxonomists and the loss of the knowledge after the existing ones retire are a great concern [18]. The inaccessibility of certain data, even in the online world, is emphasized. But this can be eliminated by an integrated approach of classical taxonomy with all the other available techniques, from biochemical assays to bioinformatics.

An interesting approach was proposed by Tautz et al. [18]. It is a DNA-based taxonomy that is a support system for mainstream taxonomy. The DNA taxonomy proposed by Tautz et al. [18] deals with sample collection, storage, deposition of data, and naming. It is an approach where a taxonomist (or a scientist or a researcher) needs to deposit sequence data (either DNA or amino acid) in a new and common database. The need for a new database rather than existing ones, such as National Center for Biotechnology Information (NCBI) or European Molecular Biology Laboratory (EMBL), arose as the existing data sets are not trusted for their taxonomical features. A uniform system to identify and deposit sequences along with the naming pattern was suggested. All in all there are advocates for DNA-based taxonomy as an alternative to classical taxonomy and others who think that the new theme might lead to "catastrophes" [19, 20].

In our view, it is evident that the cryptic species, which are almost identical in morphology and to an extent in behavior and biology, can only be separated using biochemical assays. The DNA-based assay needs fewer samples (e.g., the leg of a mosquito or a sand fly thorax is enough) and can be done on a large scale in a simple manner (loop-mediated isothermal amplification [LAMP], a simplified form of PCR). LAMP is much like PCR, except the fact that it uses a single temperature (isothermal) and more than three primers. This method holds a good future for rapid and relatively cheap diagnosis of vector species or disease in low-income countries. In addition to the isothermal nature, the results can also be readily seen with the naked eye in the form of turbidity or fluorescence. The advantage of using DNA-based assays makes them one of the prominent auxiliary tools, along with classical taxonomy.

4.2.2 Protein-Based Assays

Proteins are useful tools for assessing the differences among sibling species. These have been used for many years with various degrees of success and criticism. The changes in critical compounds such as gut receptors and salivary protein are often targeted for discriminating sibling species. Isoenzymes, or isozymes, were also used in the sibling species analysis [21].

Isozymes have been used for discriminating sibling species in some mosquitoes. They are quite easy to separate and isolate. Electrophoresis (starch gel electrophoresis or polyacrylamide gel electrophoresis) is used to separate the isozymes. The electrophoretic data were statistically analyzed using software such as Biosys [22]. But there are limitations in applying them: the need for expertise, the cost, and the fact that the sample should be more or less fresh or stored at −70°C. The method is also limited by the fact that not all isozymes can be distinguished from one another electrophoretically, leading to underdetection of variation.

4.2.3 Karyotyping and Chromosome Banding Pattern

This is one of the most promising types of techniques for discriminating sibling species. Hybridization and cross-breeding techniques (laboratory colony mating) were the predominant techniques used to identify the reproductive isolation of sibling species. But because of the limitations in the technique—crucially the laboratory environment does not fully represent the wild—often only postmating reproductive isolation is detected [21].

Karyotyping is assessing the structural variations in the chromosomes of cells in the metaphase of mitosis. The technique again requires expertise in preparation as well as analysis. Either ovarian nurse cells or the brain cells of third instar larvae of mosquitoes are used. The polytene chromosome banding pattern is used in discriminating members of species complexes. Well documented are anopheline species complexes [23]. The methods rely on there being sufficient karyotypic or chromosome banding pattern differences so as to be clearly visible with the methods available (Fig. 4.3).

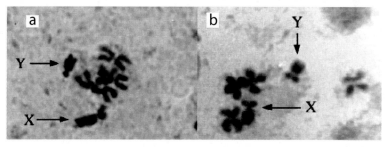

Figure 4.3   Mitotic chromosome of *A. culicifacies* sensu lato. (a) Species B with an acrocentric Y chromosome (×1000). (b) Species E with a submetacentric sex chromosome (×1000). Reprinted from Ref. [11] with permission from John Wiley and Sons, copyright 2008.

4.3   Automated Species Identification

It is a taxonomist's dream to come up with an automated species identifier that circumvents a daily manual routine of looking through the microscope for a long time to find a very minute variation such as a setae (small hairs on the body and appendages in insects) structure or a cibarial teeth number. The tediousness can often lead to inaccurate assessment. The workload itself might be responsible for some premature conclusion about a new structural variation, which can often be a result of an environmental vigor. Every year at least two new species of sand flies are reported from the Indian subcontinent. This arises from the current high level of interest in sand flies as vectors of leishmaniasis. However, the results are often single- or two-character variations from the closest species of sand flies.

What is the principle of an automated species identifier? It is simply a semiautomated or fully automated computer-aided taxonomy based on analyzing images or media files. The semiautomated system allows the interaction of users, whereas the fully automated system does not. Considering the vast improvement in computational fields and advances in bioinformatics, it is always good to look for new means and ways to use this technology to the advantage of taxonomy.

In any case the system requires an expert to first identify and calibrate the system. It is used to identify a specimen as being, or not being, similar to the type/test specimen. So it is a tool to reduce the workload of repeated identification of the same species. When one finds a new specimen that is different from the type/test specimen, much classical taxonomy work lies ahead. This is highlighted by Gaston and O'Neill [24]. They argued that the repeated common species identification might distract the taxonomists from attending more crucial tasks in taxonomy.

Despite the described taxa and the commendable work by many taxonomists, it is true that the expertise in taxonomy is fast diminishing and unevenly distributed in a global sense [25, 26]. So when an expert in the field retires or when assistance is needed from a remote locality it is always good to have a tool that can be easily used with preloaded data (image) of the described specimen. Then the user himself/herself can easily match his/her specimen using the automated species identifier, at least for common species.

Moreover there is always a need for routine identification of common species, considering improved sampling techniques and exploration of new fields. For example, consider the assessments done by environmental authorities to assess a project. They often want to describe the species in similar places. Or else consider the study of an environment/habitat in a long time period [27, 28]. This often demands a common species and a routine identification process.

Computer-assisted taxonomy (CAT) [29, 30] not only refers to automated identification but also deals with producing computer-based taxonomic tools, such as pictorial keys and resources online. CAT is often criticized for being too difficult for a general taxonomist (see Ref. [24]). But a few systems are now available for automated species recognition, for example, digital automated identification system (DAISY) [31, 32]. The automatic leafhopper identification system (ALIS) [33] is another established system for leafhopper identification. The automated bee identification system (ABIS) [34, 35]) is a tool to identify bee species.

The systems use various traits. The sound/audio signals given by various groups of animals, especially birds and insects, are again a clue for the automated identification system. The general process involved in an automated image-identifying system is taken from the text-mining protocols, such as the bag-of-words approach. In the state-of-the-art visual object recognition systems the bag-of-words model has shown excellent categorization performance [36], including large evaluations such as PASCAL VOC Challenges [37] and IMAGENET Large Scale Visual Recognition Challenge (http://image-net.org/challenges/LSVRC/2013).

The bag-of-words approach was originally used in text mining [38] and is now widely used in the retrieval of objects from a movie [39], visual object classification [40, 41], image scene classification [42, 43], and insect pest classification [44] tasks in computer vision.

The bag of words in computer vision is normally referred to as a *bag of features* (BOF) or a *bag of keypoints*. The pseudocode of the BOF approach is given in the following algorithm:

Algorithm: Process of building a BOF representation for images
for all image do
    interest Pts ← detectPts(image)
    descriptors ← describePts(interestPts)
end for
    codebook ← quantizePts(descriptors(training images))
for all image do
    BOF ← computeHistogram(codebook, descriptors(image))
end for

Interest points or regions are detected in training images and a visual codebook is then constructed by a vector quantization technique that groups similar features together.

Each group is represented by the learned cluster centers referred to as *visual words* or *codewords*. The size of the codebook is the number of clusters obtained from the clustering technique. Each interest keypoint of an image in the data set is then quantized to its closest code word in the codebook such that it maps the entire patches of an image into a fixed-length feature vector of frequency histograms, that is, the visual codebook model treats an image as a distribution of local features. Figure 4.4 shows the generic framework of such a codebook model.

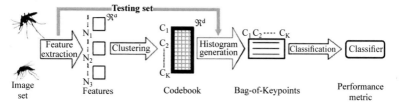

Figure 4.4 The framework of the process involved in a general bag-of-features approach used in the identification of taxa.

On the other hand the overall framework can be summarized in the following steps:

- **Feature extraction:** Detecting and describing image patches from the image corpus.
- **Cluster analysis:** Constructing visual codebooks by means of clustering techniques. Codebooks are defined as the centers of the learned clusters.
- **Histogram generation:** Mapping the extracted image descriptors into a feature vector by computing the frequency histograms with the learned clusters. This mapping produces a BOF representation.
- **Classification:** Classifying the test set to predict which category to assign to the image.

Naturally, this framework ignores the spatial layout of features corresponding to overall shapes and sizes of objects, a limitation that will require community-wide attention in the future that is outside the scope of this section. Following the great success of the BOF approach in visual object recognition, the overall framework has been then extended to biologically inspired classification schemes, such as *Drosophila* gene expression pattern annotation [45], stoneflies [46], greenhouse pests [47], and tea pests [48]. Recently Venugoban and Ramanan [44] used an approach that is a concatenation of gradient-based features, such as scale-invariant feature transform (SIFT), speeded up robust features (SURF) [49], and histogram-oriented gradient (HOG) [50], to identify paddy field insect pests in Jaffna, Sri Lanka.

As in any taxonomic scheme the automated system also possesses errors and disadvantages [24]. It is difficult to assume it as an alternative to classical taxonomy. But with careful considerations and calibrations it can be used as an auxiliary tool to mainstream taxonomy.

4.4 Species Complex among Insect Vectors in Sri Lanka: Two Case Studies Where Bioinformatic Approaches Have Solved Taxonomic Problems

The major vector-borne diseases that have been widespread in Sri Lanka are malaria, lymphaticfilariasis, dengue, and leishmaniasis. We considered malaria and leishmaniasis vectors, which were studied in detail by the authors, but it seems likely that the taxonomic concepts discussed here will apply to many other species complexes reported from different parts of the world.

Sri Lanka, where the threat of malaria has remained for years, is an island in the Indian Ocean and lies closer to the southern extremity of India. Early documents indicate that malaria may in fact have been present in Sri Lanka as early as AD 1300 [51], when a disease with malarial symptoms hit the then capital Anuradhapura. More recent outbreaks have led to the disease placing a high social and economic burden on Sri Lanka [52].

On the other hand leishmaniasis emerges as an important public health concern in Sri Lanka. The disease had been regarded to be an exotic disease among the migrant workers returning from Middle Eastern countries. The first case of autochthonous cutaneous leishmaniasis was reported in Hambantota district in 1992 [53], but the disease is now known to be more prevalent and reported from most parts of the country. Fatal visceral leishmaniasis has also been observed recently, in addition to cutaneous leishmaniasis [54]. The etiological agent causing cutaneous leishmaniasis in Sri Lanka is *Leishmania donovani* zymodeme MON 37 [55]. The interesting fact is that the donovani group of *Leishmania* is generally responsible for visceral leishmaniasis. *L. donovani* can cause dermal blisters in patients who have been affected by visceral leishmaniasis in a condition known as post-kala-azar dermal leishmaniasis (PKDL), but the exact mechanisms that drive these changes in the parasite are not well established.

The need for accurately identifying vector species is important in controlling the disease, because the most effective control strategy for many insect vector–borne diseases is control of the vector. The boundaries between species are difficult to establish where the species are very closely related to each other, such as within a species complex. Morphology, which is always based on visual clues, was historically one of the main ways for classifying organisms. Mayr's [56] biological species concept, mainly based on reproductive capability, changed the morphological approach.

We can cite the disparities caused by morphology-based taxonomy and the use of bioinformatics with a few examples where minimal variation in morphology is produced during the course of evolution and thus they share the same environment and are very closely related in their biology, too. The first one is *A. subpictus* sensu lato. The species complex plays an important role in transmitting *Plasmodium vivax* and *P. falciparam* in many parts of Sri Lanka [57–60]. The complex was reported to consist of four sibling species, namely A, B, C, and D, in India. The four members of this complex have been described as possessing characteristic paracentric fixed inversions on the X chromosome, viz. species A (X + a, +b), species B (Xa, b), species C (Xa, +b), and species D (X + a, b) [61, 62]. Additionally, stage-specific morphometric characteristics, for example, the number of ridges in egg floats, larval mesothoracic seta 4, pupal setae, and ornamentation of the palpi of adult females, have been reported to be useful for differentiating *A. subpictus* sibling species in field studies [61]. Early analysis based on a single inversion of the X chromosome and later based on the egg ridges revealed the presence of all four sibling species in Sri Lanka [63]. They have a different breeding ecology as well. A microgenomic assay targeting the D3 and ITS2 regions of the ribosomal DNA suggests that the morphology-based identification scheme proposed for the sibling species of *A. subpictus* cannot be supported by DNA sequence analysis (Figs. 4.5 and 4.6). The *A. subpictus* sibling species B identified on the basis of morphology alone was most closely related genetically to *A.*

sundaicus, which is a brackish-water malaria vector well known in Southeast Asia and not previously known to occur in Sri Lanka [64].

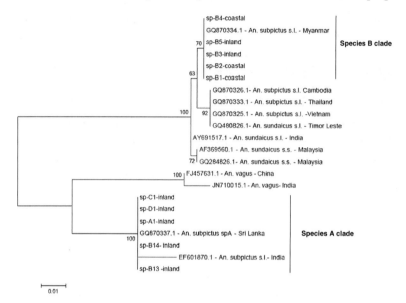

Figure 4.5 Phylogenetic analysis based on ITS2 sequences showing Sri Lankan samples claded into species A and species B. The specimens used for analysis include morphologically identified species A, B, C, and D of Sri Lanka. Reprinted from Ref. [17], copyright 2013 Surendran et al.

The major malaria vector *A. culicifacies* is reported to exist as a species complex comprising two sibling species, namely B and E, in Sri Lanka [65], with different biological properties [66]. As reported for Indian B and E [67] both sibling species were differentiated on the basis of Y chromosome (karyotype) morphology. The Y chromosome morphology was later associated with DNA sequence variations for the Indian species B and E [68]. However, Sri Lankan species B and E shared sequence similarity in many loci, including those that were used to differentiate Indian species B and E [69]. This shows karyotyping is not always useful in differentiating sibling species, and supports the involvement of bioinformatic tools.

The inadequacy of morphology-based taxonomy was also experienced in our study of sand fly vectors of the *P. argentipes* species complex. The morphometric data from India suggested the presence of at least three sibling species [70]. Later this was also confirmed in Sri Lanka [71, 72].

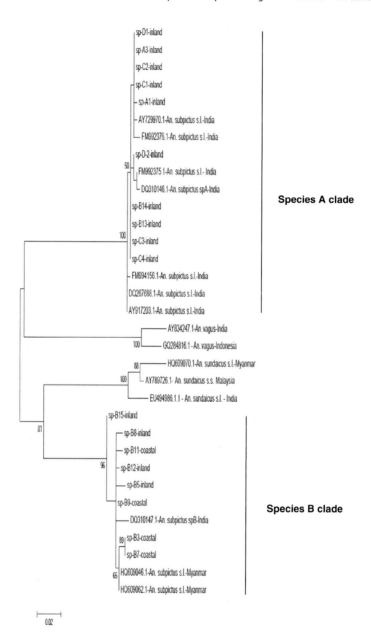

Figure 4.6 Phylogenetic analysis based on the CO1 sequence showing Sri Lankan samples claded into species A and species B. The specimens used for analysis include morphologically identified species A, B, C, and D of Sri Lanka. Reprinted from Ref. [17], copyright 2013 Surendran et al.

The only visible variations occur among the female morphospecies and are very subtle and minute. For example, the antennal ascoid or the antennal flagellomere (a very transparent sensory hair) has been identified to vary in length among them (Fig. 4.7).

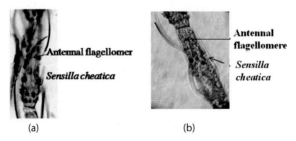

(a)   (b)

Figure 4.7   The sensilla cheatica (SC)/antennal ascoid in the second antennal flagellomere (AF) among the proposed four morphospecies of *P. argentipes* sensu lato, as seen in (a) *P. annandalei* and *P. argentipes* s.s., (b) *P. glaucus* and *P. glaucus* variants.

The variation was very minimal in another microscopic structure, known as pharyngeal armature, which refers to chitinous teeth/ scales found inside the pharynx (Fig. 4.8).

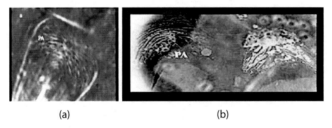

(a)   (b)

Figure 4.8   Pharyngeal armature (PA) pattern among the proposed four morphospecies of *P. argentipes* sensu lato, as seen in (a) *P. annandalei*, (b) *P. glaucus* and *P. glaucus* variants.

A similar observation was made in the maxillary teeth number as well, where there was a very slight change in the numbers of the lateral and ventral teeth found in the maxillary blade of these morphospecies (Fig. 4.9).

A principal component analysis (PCA) approach was selected in all three studies. The selected morphometric and meristic characters were analyzed on the basis of the principal components (Fig. 4.10). The score plot between the first and second principal components was clearly indicating four different groups of flies in females and males. The morphometrics analyzed in the study were chosen on the basis of their significance in taxonomy.

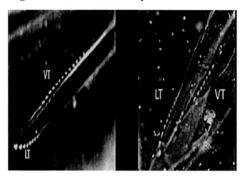

Figure 4.9 The maxillary teeth (VT: ventral teeth; LT: lateral teeth) in (a) *P. annandalei*, (b) *P. glaucus* and *P. glaucus* variants.

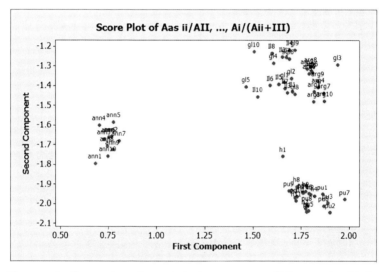

Figure 4.10 Score plots for principal component (PC) 1 and PC2 for all measured characters of females. ann: *P. annandalei*; ar: *P. argentipes* s.s.; gl: *P. glaucus*; h: Hambantota flies; pu: Punguduteevu flies; •: data points.

But the interaction with the various environments and the availability of nutrients and space may tend to shape the morphology differently among the same species. So, morphology-alone taxonomy may tend to overrepresent the diversity. This was true in the above cases.

Later it was found that only two possible closely related sibling species might exist in the *P. argentipes* complex [12]. A microgenomics approach was selected and many gene fragments such as D3, ITS2 of ribosomal DNA, and cytochrome oxidase I and cytochrome b of mitochondrial DNA were targeted to find the best cladogram (Figs. 4.11 and 4.12).

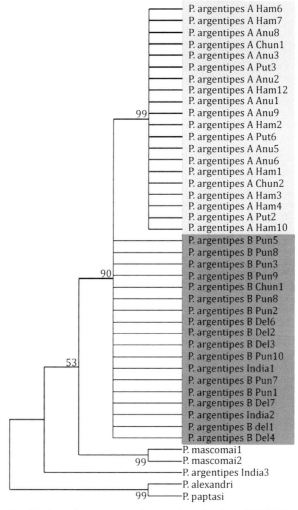

Figure 4.11 The best-fit tree created using the mitochondrial DNA sequence (cytochrome oxidase I gene) to find out the number of genetic species/sibling species found among *P. argentipes* sensu lato. Reprinted from Ref. [12], copyright 2013 Gajapathy et al.

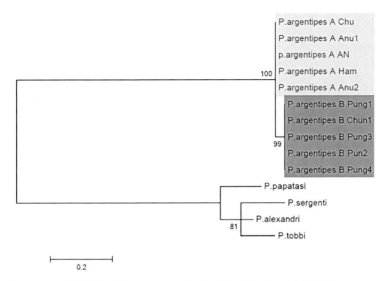

Figure 4.12 The best-fit tree created using the mitochondrial DNA sequence (cytochrome b gene) to find out the number of genetic species/sibling species found among *P. argentipes* sensu lato. Reprinted from Ref. [12], copyright 2013 Gajapathy et al.

4.5 Conclusion

The above two examples clearly indicate the potential of bioinformatics to discriminate sibling species and to assess their real taxonomic status. Along with this we can also use other computational tools, such as image analysis (although it is quite unlikely to distinguish between sibling species, it is still useful to assess any differences in gross morphology to identify any distinguishing features) and processing, to achieve a more reliable way to tackle the current problems associated with taxonomy.

One area that these studies illustrate clearly is the difficulty in establishing where intraspecific polymorphism ends and where new, discrete species begin, particularly in absence of data on reproductive compatibility among the genetic variants identified. This is particularly problematic if species are only recently diverged from one another and share many polymorphisms that evolved prior to the speciation event. In this regard it may again be that a bioinformatics approach provides a useful way forward: next-generation sequencing (NGS) describes a suite of techniques by which RNA and DNA sequences from large parts of (or entire) genomes and transcriptomes can be generated from hitherto uncharacterized species (reviewed in Ref. [73]). A wealth of data already exists for many closely related species pairs. These data sets are used to detect those small changes that likely determine reproductive compatibility and species identity. *Heliconius* butterflies are one well-studied invertebrate example where bioinformatic analysis of molecular genetic data has provided insights into the relationships among closely related cryptic species that are mimics of each other [74]. Full-scale genome sequencing using the NGS approach is now being applied to a wide range of species within the group. The potential to identify a small number of genetic changes that differ between species not only identifies species diagnostic markers but ultimately also gives insights into the speciation process itself. Beyond the application of bioinformatic tools to identify and distinguish among insects, there are many other groups of organisms where a similar approach would be beneficial. Spiders are one such example; in many cases individuals cannot be identified to species level or sexed until they have reached—or are close to reaching—maturity. Molecular techniques provide an alternative method for determining species identity in cases where morphology is not enough. The advent of technology that allows sequencing and analysis of large sections of the genome will likely render the approach of sequencing small sections of mitochondrial DNA (molecular "barcoding") obsolete, with species assignment relying upon much larger quantities of DNA sequences (reviewed in Ref. [73]). Linyphiid spiders, such as the agrobiont species that are commonly associated with farmed landscapes, are a good illustration not only of the problems encountered when trying to identify individuals to species level but also the benefits of being able to do so. In the sand fly and mosquito examples already described, species identification can further our understanding of disease transmission. In the case of linyphiid spiders it furthers our understanding of the population dynamics of species with the capacity to eat agricultural pests, thereby reducing the need for insecticides [75]. Distinguishing between venomous and nonvenomous species within other groups of spiders in cases where morphological similarities are high is another example of the benefit of being able to accurately identify species.

Bioinformatic tools are likely to provide increasingly powerful ways of assessing levels of biodiversity and distinguishing between morphologically similar species. Furthermore, combinations of genetic and ecological data can allow inferences to be made about the processes that generated these morphologically indistinguishable species.

References

1. Knowlton, N. (1986). Cryptic and sibling species among the decapods Crustacea. *J. Crustac. Biol.*, **6**, pp. 356–363.

2. Winker, K. (2005). Sibling species were first recognized by William Derham (1718). *Auk*, **122**, pp. 706–707.

3. Bickford, D., Lohman, D. J., Navjot, S. S., Ng, P. K. L., Meier, R., Winker, K., Ingram, K. K., and Das, I. (2007). Cryptic species as a window on diversity and conservation. *Trends. Ecol. Evol.*, **22**, pp. 148–155.

4. Bowen, B. W., Nelson, W. S., and Avise, J. C. (1993). A molecular phylogeny for marine turtles: trait mapping, rate assessment, and conservation relevance. *Proc. Natl. Acad. Sci. U. S. A.*, **90**, pp. 5574–5577.

5. Ravaoarimanana, I. B., Tiedemann, R., Montagnon, D., and Rumpler, Y. (2004). Molecular and cytogenetic evidence for cryptic speciation within a rare endemic Malagasy lemur, the Northern Sportive Lemur (*Lepilemur septentrionalis*). *Mol. Phylogenet. Evol.*, **31**, pp. 440–448.

6. Meegaskumbura, M., Bossuyt, F., Pethiyagoda, R., Manamendra-Arachchi, K., Bahir, M., Milinkovitch, M. C., and Schneider, C. J. (2002). Sri Lanka: an amphibian hot spot. *Science*, **298**, p. 379.

7. Grundt, H. H., Kjølner, S., Borgen, L., Rieseberg, L. H., and Brochmann, C. (2006) High biological species diversity in the arctic flora. *Proc. Natl. Acad. Sci. U. S. A.*, **103**, pp. 972–975.

8. Vrijenhoek, R. C., Schutz, S. J., Gustafson, R. G., and Lutz, R. A. (1994). Cryptic species of deep-sea clams (Mollusca, Bivalvia, Vesicomyidae) in hydrothermal vent and cold-seep environments. *Deep-Sea Res. II*, **41**, pp. 1171–1189.

9. Davison, A., and Clarke, B. C. (2000). History or current selection? A molecular study of area effects in *Cepaeanemoralis. Proc. R. Soc. London B*, **267**, pp. 1399–1405.

10. Subbarao, S. K., Adak, T., Vasantha, K., Joshi, H., Raghavendra, K., Cochrane, A. H., Nussenzweig, R. S., and Sharma V. P. (1988).

Susceptibility of *Anopheles culicifacies* species A and B to *Plasmodium vivax* and *Plasmodium falciparum* as determined by immune radiometric assay. *Trans. R. Soc. Trop. Med. Hyg.*, **82**, pp. 394–397.

11. Surendran, S. N., Abhayawardana, T. A., de Silva, B. G. D. N. K., Ramasamy, M. S., and Ramasamy, R. (2000). *Anopheles culicifacies* Y-chromosome dimorphism indicates the presence of sibling species (B and E) with different malaria vector potential in Sri Lanka. *Med. Vet. Entomol.*, **14**, pp. 437–440.

12. Gajapathy, K., Peiris, L. B. S., Goodacre, S. L., Silva, A., Jude, P. J., and Surendran, S. N. (2013). Molecular identification of potential leishmaniasis vector species within the *Phlebotomus (Euphlebotomus) argentipes* species complex in Sri Lanka. *Parasit. Vectors*, **6**, p. 302.

13. Hammond, P. M. (1992). Species inventory. In *Global Biodiversity: Status of the Earth's Living Resources*, pp. 17–39, Groombridge, B. (ed.) (Chapman and Hall, London).

14. Hawksworth, D. L., and Kalin-Arroyo, M. T. (1995). Magnitude and distribution of biodiversity. In *Global Biodiversity Assessment*, pp. 107–199, Heywood, V. H. (ed.) (Cambridge University Press).

15. Hebert, P. D. N., Cywinska, A., Ball, S. L., and de Waard, J. R. (2003). Biological identifications through DNA barcodes. *Proc. R. Soc. London B*, **270**, pp. 313–321.

16. Will, K. W., and Rubinoff, D. (2004). Myth of the molecule: DNA barcodes for species cannot replace morphology for identification and classification, *Cladistics*, **20**, pp. 47–55.

17. Surendran, S. N., Sarma, D. K., Jude, P. J., Kemppainen, P., Kanthakumaran, N., Gajapathy, K., Peiris, L. B. S., Ramasamy, R., and Walton, C. (2013). Molecular characterization and identification of members of the *Anopheles subpictus* complex in Sri Lanka. *Malar. J.*, **12**, p. 304.

18. Tautz, D., Arctander, P., Minelli, A., Thomas, R. H., and Vogler, A. P. (2003). A plea for DNA taxonomy. *Trends Ecol. Evol.*, **18**, pp. 71–74.

19. Rubinoff, D. (2006). Utility of mitochondrial DNA barcodes in species conservation. *Conserv. Biol.*, **20**(4), pp. 1026–1033.

20. Will, K. W., Mishler, B. D., and Wheeler, Q. D. (2005). The perils of DNA bar-coding and the need for integrative taxonomy, *Syst. Biol.*, **54**, pp. 844–851.

21. Naranag, S. K., and Seawright, J. A. (1998). Electrophoretic methods for recognizing sibling species of Anopheline mosquitoes a practical approach. *Fla. Entomol.*, 71, pp. 303–311.

22. Swoffird, D. L., and Selander, R. B. (1981). Biosys-1 A FORTRAN programme for comprehensive analysis for electrophoretic data in population genetics and systematic. *J. Hered.*, **72**, pp. 281–283.

23. World Health Organization (2008). *Anopheline Species Complexes in South and South-East Asia*. SEARO Technical Publications No. 57.

24. Gaston K. J., and O'Neill M. A. (2004). Automated species identification: why not? *Phil. Trans. R. Soc. London B.*, **359**, pp. 655–667.

25. Simonetti, J. A. (1997). Biodiversity and a taxonomy of Chilean taxonomists. *Biodiv. Conserv.*, **6**, pp. 633–637.

26. Hopkins, G. W., and Freckleton, R. P. (2002). Decline in the numbers of amateur and professional taxonomists: implications for conservation. *Anim. Conserv.*, **5**, pp. 245–249.

27. Culverhouse, P. F., Simpson, R. G., Ellis, R., Lindley, J. A., Williams, R., Parisini, T., Reguera, B., Bravo, I., Zoppoli, R., Earnshaw, G., McCall, H., and Smith, G. (1996). Automatic classification of field-collected dinoflagellates by artificial neural network. *Mar. Ecol. Prog. Ser.*, **139**, pp. 281–287.

28. France, I., Duller, A. W. G., Duller, G. A. T., and Lamb, H. F. (2000). A new approach to automated pollen analysis. *Quatern. Sci. Rev.*, **19**, pp. 537–546.

29. Chesmore, D. (1999). Technology transfer: applications of electronic technology in ecology and entomology for species identification. *Nat. Hist. Res.*, **5**, pp. 111–126.

30. Chesmore, D. (2000). Methodologies for automating the identification of species. In *Proceedings of the BioNET INTERNATIONAL Group for Computer-Aided Taxonomy (BIGCAT)*, pp. 3–12, Chesmore, D., Yorke, L., Bridge, P., and Gallagher, S. (eds.).

31. Gauld, I. D., O'Neill, M. A., and Gaston, K. J. (2000). Driving Miss Daisy: the performance of an automated insect identification system. In *Hymenoptera: Evolution, Biodiversity and Biological Control*, pp. 303 312, Austin, A. D., and Dowton, M. (eds.) (Collingwood, VIC: CSIRO).

32. O'Neill, M. A., Gauld, I. D., Gaston, K. J., and Weeks, P. J. D. (2000). Daisy: an automated invertebrate identification system using holistic vision techniques. In *Proceedings of the Inaugural Meeting of the BioNET INTERNATIONAL Group for Computer-Aided Taxonomy (BIGCAT)*, pp. 13–22, Chesmore, D., Yorke, L., Bridge, P., and Gallagher, S. (eds.).

33. Dietrich, C. H., and Pooley, C. D. (1994). Automated identification of leafhoppers (Homoptera: Cicadellidae). *Ann. Entomol. Soc. Am.*, **87**, pp. 412–423.

34. Arbuckle, T., Schroder, S., Steinhage, V., and Wittmann, D. (2001). Biodiversity informatics in action: identification and monitoring of bee species using ABIS. In *Proceedings of the 15th International Symposium on Informatics for Environmental Protection*, ETH Zurich, 10–12 October 2001, pp. 425–430.

35. Arbuckle, T. (2002). Automatic identification of bees' species from images of their wings. In *Proceedings of the 9th International Workshop on Systems, Signal and Image Process*ing, UMIST, Manchester, pp. 509–511.

36. Ramanan, A., and Niranjan, M. (2011). A review of codebook models in patch-based visual object recognition. *J. Sign. Process. Syst.*, **68**, pp. 333–352.

37. Everingham, M., Van Gool, L., Williams, C. K. I., Winn, J., and Zisserman, A. (2010). The Pascal visual object classes (VOC) challenge. *Int. J. Comp. Vis. (IJCV)*, **88**, pp. 303–338.

38. Joachims, T. (1998). Text categorization with support vector machines: learning with many relevant features. *Proc. Eur. Conf. Mach. Learn. (ECML'98)*, pp. 137–142.

39. Sivic, J., and Zisserman, A. (2003). Video Google: a text retrieval approach to object matching in videos. *Proc. Ninth IEEE Int. Conf. Comp. Vis. (ICCV'03)*, pp. 1470–1478.

40. Csurka, G., Dance, C. R., Fan, L. Willamowski, J., and Bray, C. (2004). Visual categorization with bags of keypoints. In *Workshop Stat. Learn. Comp. Vis. (ECCV'04)*, pp. 1–22.

41. Zhang, W., Surve, A., Fern, X., and Dietterich, T. (2009). Learning non-redundant codebooks for classifying complex objects. *Proc. 26th Int. Conf. Mach. Learn. (ICML'09)*.

42. Fei-Fei, L., and Perona, P. (2005). A Bayesian hierarchical model for learning natural scene categories. *Proc. IEEE Conf. Comp. Vis. Pattern Recog. (CVPR'05)*, **2**, pp. 524–531.

43. Quelhas, P., Monay, F., Odobez, J. M., Gatica-Perez, D., and Tuytelaars, T. (2007). A Thousand words in a scene. *IEEE Trans. Pattern Anal. Mach. Intell.*, **29**, pp. 1575–1589.

44. Venugoban, K., and Ramanan, A. (2014). Image classification of paddy field insect pests using gradient-based features. *Int. J. Mach. Learn. Comput. (IJMLC)*, **4**, pp. 1–5.

45. Ji, S., Li, Y.-X., Zhou, Z.-H., Kumar, S., and Ye, J. (2009). A bag-of-words approach for Drosophila gene expression pattern annotation. *BMC Bioinf.*, **10**, p. 119.

46. Larios, N., Deng, H., Zhang, W., Sarpola, M., Yuen, J., Paasch, R., Moldenke, A., Lytle, D. A., Correa, S. R., Mortensen, E., Shapiro L., and Dietterich, T. (2008). Automated insect identification through concatenated histograms of local appearance features: feature vector generation and region detection for deformable objects. *Mach. Vis. Appl.*, **19**, pp. 105–123.

47. Mundada, R. G., and Gohokar, V. V. (2013). Detection and classification of pests in greenhouse using image processing. *IOSR J. Electron. Commun. Eng.*, **5**, pp. 57–63.

48. Samanta, R. K., and Ghosh, I. (2012). Tea insect pests classification based on artificial neural networks. *Int. J. Comp. Eng. Sci. (IJCES)*, **2**, pp. 1–13.

49. Lowe, D. G. (1999). Object recognition from local scale-invariant features. *Proc. Int. Conf. Comp. Vis.*, **2**, pp. 1150–1157.

50. Dalal, N., and Triggs, B. (2005). Histograms of oriented gradients for human detection. *Proc. IEEE Conf. Comp. Vis. Pattern Recog.*, **1**, pp. 886–893.

51. Karunaratne, W. A. (1959). The influence of malaria control on vital statistics in Ceylon. *J. Trop. Med. Hyg.*, **62**, pp. 79–85.

52. Sri Lanka Ministry of Health. (2008). *Anti-Malaria Campaign: Strategic Plan for Phased Elimination of Malaria 2008–2012*, Health Ministry, Sri Lanka.

53. Athukorale, D., Seneviratne, J., Ihalamulla, R., and Premaratne, U. (1992). Locally acquired cutaneous leishmaniasis in Sri Lanka. *J. Trop. Med. Hyg.*, **95**, pp. 432–433.

54. Siriwardana, H., Thalagala, N., and Karaunaweera, N. (2010). Clinical and epidemiological studies on the cutaneous leishmaniasis caused by *Leishmania (Leishmania) donovani* in Sri Lanka. *Ann. Trop. Med. Parasit.*, **104**, pp. 213–223.

55. Navaratna, S. S. K., Weligama, D. J., Wijekoon, C. J., Dissanayake M., and Rajapaksha, K. (2007). Cutaneous leishmaniasis, Sri Lanka. *J. Emerg. Inf. Dis.*, **13**, pp. 1068–1070.

56. Mayr, E. (1963). *Animal Species and Evolution* (Belknap Press of Harvard University Press, Cambridge).

57. Amerasinghe, F. P., and Ariyasena, T. G. (1990). Larval survey of surface water-breeding mosquitoes during irrigation development in the Mahaweli project, Sri Lanka. *J. Med. Entomol.*, **27**, pp. 789–802.

58. Amerasinghe, F. P. (1992). A guide to the identification of the anopheline mosquitoes (Diptera: culicidae) of *Sri Lanka*: II. Larvae. *Ceylon J. Sci. (Biol. Sci.)*, **22**, pp. 1–13.

59. Ramasamy. R., de Alwis, R., Wijesundere, A., and Ramasamy, M. S. (1992). Malaria transmission at a new irrigation project in Sri Lanka: the emergence of *Anopheles annularis* as a major vector. *Am. J. Trop. Med. Hyg.*, **47**, pp. 547–553.

60. Surendran, S. N., and, Ramasamy, R. (2010). The *Anopheles culicifacies* and *An. Subpictus* complexes in Sri Lanka and their implications for malaria control in the country. *Trop. Med. Health*, **38**, pp. 1–11.

61. Suguna, S. G., Rathinam, K. G., Rajavel, A. R., and Dhanda, V. (1994). Morphological and chromosomal descriptions of new species in the *Anopheles subpictus* complex. *Med. Vet. Entomol.*, **8**, pp. 88–94.

62. Chandra, G., Bhattacharjee, I., and Chatterjee, S. (2010). A review on *Anopheles subpictus* Grassi, a biological vector. *Acta. Trop.*, **115**, pp. 142–154.

63. Abhayawardana, T. A., Wijesuriya, S. R., and Dilrukshi, R. K. (1996). *Anopheles subpictus*complex: distribution of sibling species in Sri Lanka. *Indian J. Malariol.*, **33**, pp. 53–60.

64. Surendran, S. N., Jude, P. J., Singh, O. P., and Ramasamy, R. (2010). Genetic evidence for the presence of malaria vectors of *Anopheles sundaicus* complex in Sri Lanka with morphological characteristics attributed to *Anopheles subpictus* species B. *Malar. J.*, **9**, p. 343.

65. Surendran, S. N., Abhayawardana, T. A., de Silva, B. G. D. N. K., Ramasamy, M. S., and Ramasamy, R. (2000). *Anopheles culicifacies* Y chromosome dimorphism indicates the presence of sibling species (B and E) with different malaria vector potential in Sri Lanka. *Med. Vet. Entomol.*, **14**, pp. 437–440.

66. Surendran, S. N., Ramasamy, M. S., de Silva, B. G. D. N. K., and Ramasamy, R. (2006). *Anopheles culicifacies* sibling species B and E in Sri Lanka differ in longevity and in their susceptibility to malaria parasite infection and common insecticide. *Med. Vet. Entomol.*, **20**, pp. 153–156.

67. Kar, I., Subbarao, S. K., Eapen, A., Ravindran, J., Satyanarayana, T. S., Raghavendra, K., Nanda, N., and Sharma, V. P. (1999). Evidence for a new vector species E within the *Anopheles culicifacies* complex (Diptera: culicidae). *J. Med. Entomol.*, **36**, pp. 595–600.

68. Goswami, G., Raghavendra, K., Nanda, N., Gakhar, S. K., and Subbarao, S. K. (2006). PCR-RFLP of mitochondrial cyctochrome oxidase sub unit II and ITS2 ribosomal DA: markers for the identification of members of the *Anopheles culicifacies* complex (Diptera: culicidae). *Acta. Trop.*, **95**, pp. 92–99.

69. Surendran, S. N., Hawkes, N. J., Steven, A., Hemingway, J., and Ramasamy, R. (2006). Molecular studies of *Anopheles culicifacies* (Diptera: culicidae) in Sri Lanka: sibling species B and E show sequence identity at multiple loci. *Eur. J. Entomol.*, **103**, pp. 233–237.

70. Ilango, K. (2010). A taxonomic reassessment of the *Phlebotomus argentipes* species complex (Diptera: Psychodidae: Phlebotominae). *J. Med. Entomol.*, **47**, pp. 1–15.

71. Gajapathy, K., Jude, P., and Surendran, S. N. (2011). Morphometric and meristic characterization of *Phlebotomus argentipes* species complex in northern Sri Lanka: evidence for the presence of potential leishmaniasis vectors in the country. *Trop. Biomed.*, **28**, pp. 259–268.

72. Ranasinghe, S., Maingon, R. D., Bray, D. P., Ward, R. D., Udagedara, C., Dissanayake, M., Jayasuriya, V., and de Silva, N. K. (2012). A morphologically distinct *Phlebotomus argentipes* population from active cutaneous leishmaniasis foci in central Sri Lanka. *Mem. Inst. Ozwaldo. Cruz.*, **107**, pp. 402–409.

73. Goodacre, S. L. (2013). Spider genetics and genomics. In *Spider research of the 21st Century*, pp. 184–199, Penney, D. (ed.) (Siri Press).

74. Giraldo, N., Salazar, C., Jiggins, C. J., Bermingham, E., and Linares, M. (2008). Two sisters in the same dress: *Heliconius* cryptic species. *BMC Evol. Biol.*, **8**, p. 324.

75. Thomas, C. F. G., Blackshaw, R. P., Hutchings, L., Woolley, C., Goodacre, S. L., Hewitt, G. M., Ibrahim, K., Brooks, S. P., and Harrington, R. (2003). Modeling life history/dispersal strategy interactions to predict and manage linyphiidspdier diversity in agricultural landscapes. *IOBC WPRS Bull.*, **26**, pp. 167–172.

Chapter 5

Integrated Bioinformatics, Biostatistics, and Molecular Epidemiologic Approaches to Study How the Environment and Genes Work Together to Influence the Development of Complex Chronic Diseases

Alok Deoraj,[a] Changwon Yoo,[b] and Deodutta Roy[a]

[a]*Department of Environmental and Occupational Health, Florida International University, Robert Stempel College of Public Health and Social Work, 11200 SW 8th Street, AHCII 596, Miami, FL 33199-0001, USA*
[b]*Department of Biostatistics, Florida International University, Robert Stempel College of Public Health and Social Work, 11200 SW 8th Street, AHCII 596, Miami, FL 33199-0001, USA*
droy@fiu.edu

Most of the chronic diseases occur as a result of adverse effects at multiple points on behavioral, biochemical, genetic, and physiological systems, often from multiple exposures and across various life stages. There is also a tremendous interindividual variability in the health outcome in response to environmental and

Gene–Environment Interaction Analysis: Methods in Bioinformatics and Computational Biology
Edited by Sumiko Anno
Copyright © 2016 Pan Stanford Publishing Pte. Ltd.
ISBN 978-981-4669-63-4 (Hardcover), 978-981-4669-64-1 (eBook)
www.panstanford.com

stochastic factors. This has hindered the ability of the scientific community to pinpoint why certain individuals develop a chronic disease when exposed to environmental and stochastic factors, while others remain healthy. Recent advances in bioinformatics and molecular genetic tools have provided an opportunity to understand how the genetic and epigenetic variability of an individual interacts with environmental and stochastic factors to either preserve health or cause disease. Emerging consensus indicates that susceptibility to many complex diseases is influenced by the interactions of unique inherited DNA sequences and variations in the epigenetic and consequent biochemical milieu of germ and somatic cells with environmental and stochastic factors during the intrauterine to postnatal life, childhood, and adult life of an individual. Such explanations in unrelated individuals due to low DNA sequence variation and experimental results from closely related mammalian models are inadequate to account for differences in complex chronic disease outcomes. Despite much information on both genetic and environmental disease risk factors, there are relatively few examples of reproducible gene–environment interactions (GEI). Currently, there is also a lack of computational and bioinformatics methods that can reduce large and diverse environmental, epigenetic, epidemiological, and "-omics" data sets into representations that can be interpreted in a biological context. This chapter presents an integrated bioinformatics, biostatistics, and molecular epidemiologic approach to studying the relative contributions of environmental, epigenetic, genetic, and stochastic factors in transdisciplinary molecular epidemiological studies to determine the causality and progression of complex chronic disease phenotypes.

5.1   Introduction

Complex chronic diseases with a multifactorial etiology, such as Alzheimer's, autism, diabetes, cancer, hypertension, Parkinson's, and several neurodevelopmental and mental health disorders, have become the dominant public health burden. The interplay of multiple genetic and epigenetic variations and environmental and stochastic factors influencing biological pathways and networks contributes to the susceptibility to and development of complex diseases, as well as differences in treatment responses [1]. As depicted in Fig. 5.1, it is the combined contributions or cancellations of the effects of a multitude of genetic, epigenetic, and environmental stressors and factors that lead to the development and progression of complex diseases over a period of time.

Figure 5.1   Development and progression of chronic and complex diseases is the result of combined contributions of a multitude of genetic, epigenetic, and environmental stressors and factors over a period of time.

Therefore, it is critical to design an integrated approach to include biostatistics, bioinformatics, and molecular, genetic, and epidemiologic aspects to unravel the link between the environment and health in order to understand the susceptibility and resistance to the development of complex chronic diseases. An integrated approach will also help describe the individual variations in responses to therapeutic interventions. It is realized now that a complex chronic disease outcome in an individual is the result of the collaboration of genes and environmental factors causing missing hormones and altering epigenomes of rogue cells. The interacting environment and genes that influence the origins of complex chronic diseases are also not necessarily the same as those that contribute to the chronic disease progression or, for example, metastasis of cancer. Susceptibility gene variants for each specific disease are being identified, with emerging evidence of gene–environment interactions (GEI).

Although most chronic diseases are the result of complex interactions between genes (G) and environmental (E) factors, the majority of the analytical approaches adopted for genetic linkage and association studies do not incorporate interactive effects with environmental factors. Studies have indicated that failure to account for GEI in complex chronic disease association analyses can decrease the power to find genetic disease loci, and underestimate effects of both genetic and environmental contributions in the origin and progression of complex diseases. The failure to replicate many genetic association studies is believed to be, in part, due to the omission of functional aspects of GEI in the study designs. The potential for detecting gene function(s) can be enhanced by taking environmental agents into account that are already known to play a key role in the disease etiology, particularly if these interactions are already known (e.g., tobacco smoking is a known environmental risk factor for lung or colorectal cancer). The success of genetic studies, in general, in identifying genetic variants for complex diseases will, therefore, be dependent on the further development of methods and analytical tools that can incorporate these complex functional interactions. Additionally, to detect, characterize, and interpret GEI that may cause complex diseases, innovative strategies and new tools are necessary to test multiple genes and multiple environmental risk factors, along with standard linkage analyses tools and candidate gene approaches. Limited functional tools are available to identify genetic variability and the genetic factors in an individual person, but it is difficult to define the environmental and stochastic agents that an individual is exposed to during his or her life span.

Many genetic approaches are being utilized to understand disease susceptibility. These advancements are necessitating a shift in our scientific strategies for studying risk factors for chronic and complex diseases. The complex chronic diseases are not just one disease; there are hundreds or even more. For example, cancer is not caused by one agent or one environmental factor. To develop an integrative approach to measuring the contributions of gene and environmental stressors on the chronic disease outcome, it is important to consider that during the development of an individual from a single cell to prenatal stages to adolescent to adulthood and through the complete life span, the individual is exposed to countless environmental stressors. As we know, the environment constitutes everything that surrounds us both internally and externally, including toxicants, hormones, diet, psychosocial behaviors, and lifestyles. Like genes, these factors also interact among themselves. A single exposure to an internal or external environmental factor alone cannot explain the development of a complex chronic disease; rather it appears that exposure to multiple environmental and stochastic factors across the life span and their interactions influence the development of a chronic disease in an individual. The temporal and spatial environmental modulations of the normal genetic and phenotypic changes in a cell lead to the development of a particular type of disease phenotype. In the realm of GEI, which affect disease phenotypes, toxicants such as tobacco smoke and alcohol, redox state, hormones, and diet are the best-studied environmental factors.

Traditionally, it is believed that an orderly progression of changes in disease stem cells through a recognizable series of intermediate states leads toward a predictable end point, the disease, in equilibrium with the prevailing environment [1]. In contrast, a more recent view is based on adaptations of independent disease stem cells. Transitions between a series of different states of disease stem cells are disorderly and unpredictable, resulting from probabilistic processes such as invasion, death, differentiation, and survival of disease stem cells that make up rogue tissue or a malignant lesion. This reflects the inherent variability observed in behaviors of different types of cells present in the affected tissue in their time and space and the uncertainty of environmental and stochastic factors. In particular, it allows for a succession of alternative pathways and end points dependent on the chance outcome of gene–gene interactions (G × G), environment–environment interactions, and GEI and interactions between cells and their environment. An understanding of the interactions between multiple genetic, epigenetic, environmental, and stochastic factors will more accurately predict the probability of disease risk and variations in therapeutic response [2]. The integrated GEI mechanisms will also help to better explain the development and progression of a complex disease than any correlations with a single genetic, epigenetic, stochastic, or environmental factor. The integration of genomics, proteomics, transcriptomics, and metabolomics to identify important perturbations of normal biological pathways, networks, and systems influenced by environmental factors is necessary to understand the mechanistic role of the environment in complex chronic diseases, including cancers of different origin [2, 3]. Genetic, epigenetic, epidemiologic, biostatistics, and bioinformatics approaches need to be integrated to develop study designs and analytical strategies for identifying G × G and GEI in molecular epidemiologic and genomic studies with applicability of such understanding in complex chronic diseases. This chapter presents an integrated bioinformatics, biostatistics, and molecular epidemiologic approach to studying the relative contributions of environmental, epigenetic, genetic, and stochastic factors in transdisciplinary molecular epidemiological studies to determine the causality and progression of complex chronic disease phenotypes.

5.2 Tools for Identifying Genetic, Environmental, and Stochastic Factors Relevant in Complex Human Diseases

The availability of the human genome sequence and advancement in the technologies for genomic analysis have allowed the expansion of genetic variance investigation to probe the susceptibility or resistance to complex diseases [4]. Variations in genes can occur at several genomic levels, including in single nucleotides, small stretches of DNA (microsatellites), whole genes, regulatory elements, noncoding areas, and structural components of chromosomes or complete chromosomes, which then influence G × G and GEI in response to various environmental exposures. The basis of genetic evaluation is the identification of the allelic variants of human genes. The process of identifying DNA variation that may be associated with a complex disease is continuously being catalogued and mapped. Measurement of the frequency of these DNA variants in different susceptible or resistant populations determines the magnitude of the associated risk due to their interactions with other genes and the environment in the initiation and progression of complex chronic diseases. Currently available "-omics" (proteomics, transcriptomics, interactomics, etc.) tools may aid in the identification of alterations that result in *loss or gain of functions* that contribute to the disease outcome. For example, possibly complex diseases such as cancer might be more susceptible to the lower levels, or "softer" forms, of variation, such as variation in noncoding sequences and copy number, which alter gene doses without abolishing gene function. Recently it has been shown that DNA variants associated with disease traits are concentrated in noncoding regulatory regions of the human genome that are marked by hypersensitivity to the enzyme deoxyribonuclease I (DNase I) [5]. When appropriately integrated with other molecular, cellular, physiological, and environmental data, such information may improve the way we understand normal conditions and diagnose and treat complex chronic diseases.

5.2.1 Candidate Gene

In a study design to identify candidate gene markers it is important that selection of the gene be based on (i) prior information about biological pathways or linkage data, (ii) functional correlation for a single-nucleotide polymorphism (SNP) or haplotype of the gene, including the pathway or the use of evolution-based approaches, that shows sequence homology or belongs to a gene family, and (iii) SNP haplotype studies that start with simple haplotypes, often including known nonsynonymous SNPs (nsSNPs) or regulatory SNPs [6]. Later these SNPs can be expanded to increase the density of SNPs across the haplotype. With respect to candidate genes, the paradigm for functional gene discovery began with the use of the distribution of phenotypic traits to infer genetic effects. More recently, it has been possible to relate functionally significant DNA sequence variation to clinically important variability [7]. Both of these approaches are complementary and should be done to understand the functional significance of genes and SNPs. Candidate gene expression profiling using microarray or similar technology can be valuable to identify and characterize candidate genes (e.g., for treatment response). Candidate gene expression profiles can differ by environmental exposure or treatment (i.e., where combinations of drug treatments did not evoke the same expression profile as each treatment individually). Therefore, expression profile approaches may be useful for identifying novel genes, characterizing function, classifying novel diseases, and studying GEI in the development and progression of complex diseases. Candidate gene approaches have the advantage of maximizing inferences about biological networks and complex disease causality [1, 8]. The underlying pleiotropic mechanisms of candidate gene in GEI need to be taken into consideration because they often have influence on the correlations with their effect on metabolic pathways that contribute to different phenotypes of complex diseases.

5.2.2 Copy Number Variations

Recent discoveries have revealed that large segments of DNA, ranging in size from thousands to millions of DNA bases, can vary in copy number [9, 10]. Such copy number variations (or CNVs) can encompass genes, leading to dosage imbalances. A CNV, a segment of DNA, ranging from 1 kilobase to several megabases in size, representing an imbalance between two genomes, can manifest in numerous ways, including DNA amplification and deletions and copy number gains and losses [4]. While mechanisms by which environmental exposures cause SNP mutations are well characterized, evidence of environmental exposures causing a CNV is still developing [9, 11]. Recent evidence demonstrates that many, if not most, disease-associated CNVs arise via mechanisms coupled to aberrant DNA replication and/or nonhomologous repair of DNA damage [12]. These findings also indicate that our DNA is less than 99.9% identical, as was previously thought. It is recognized then that the genomes of any two individuals in the human population differ more at the structural level than at the nucleotide sequence level [10]. This led to the discovery that structural variations of the genome, including large insertions and deletions of DNA, collectively termed CNVs, as well as balanced chromosomal rearrangements, such as inversions, contribute to a major proportion of genetic difference in humans [13].

5.2.3   Single-Nucleotide Polymorphism

Changes in a single base pair of the DNA sequence are the most frequently occurring form of variation in the human genome [6, 11]. Many genes have a large number of SNPs, and it is acknowledged that there are more than 10 million SNPs across the human genome, with an estimated two common missense variants per gene, making it impossible for cost-effective genotyping of all of them in studies of complex diseases, even in very small samples. New approaches, however, can reduce the genotyping burden by exploiting the strong correlation between some SNPs that are close together on the genome. This is due to the phenomenon of linkage disequilibrium (LD), or nonrandom association of SNP alleles at the population level, due to the sharing by multiple individuals of ancestral chromosomal segments. These segments, or haplotypes, are combinations of particular SNP alleles on the same chromosome that tend to segregate together. By choosing a subset of maximally informative SNPs, or tag SNPs, to represent these haplotypes, the number of SNPs to be genotyped in a larger sample can be reduced without losing the ability to capture most of the variation and, in particular, any association between unmeasured causal alleles and the disease outcome measured on individuals in the sample. The choice of tag SNPs becomes challenging when study subjects are from multiple populations, since the transferability of tag SNPs depends on similarity of LD patterns. It is also desirable to incorporate resequencing data from the local case and control samples generated during a SNP discovery phase. SNPinfo has combined the power of three pipelines for SNP selection on the basis of candidate genes, the whole genome, and linkage regions. Recent surveys of human genetic diversity have estimated that there are about 100,000–300,000 SNPs in protein coding sequences (cSNPs) of the entire human genome. cSNPs are of particular interest because some of them, termed "nsSNPs" or missense variants, introduce amino acid changes into their encoded proteins [6]. Alterations in the regulation of genes by regulatory SNPs may influence the kinetic parameters of enzymes, the DNA-binding properties of proteins that regulate transcription, the signal transduction activities of transmembrane receptors, and the architectural roles of structural proteins, which are all susceptible to perturbation by nsSNPs and their associated amino acid polymorphisms. Similarly, the regulation of key elements of a pathway can alter the risk for cancer and complex chronic disease outcomes either directly or through interacting pathways through environmental exposures. Identification and characterization of such regulatory variants are projected to be a major breakthrough in the future studies to map out the susceptibility and/or resistance to one or other complex chronic diseases.

5.2.4   Haplotype Mapping

Haplotype Mapping (HapMap) is a catalog of common genetic variants that occur in human beings. It describes what these variants are, where they occur in our DNA, and how they are distributed among people within populations and among populations in different parts of the world. HapMap focuses only on common SNPs, which are about 1% of the population [14]. The International HapMap Project is not using the information in the HapMap to establish connections between particular genetic variants and diseases [15]. Rather, the project is designed to provide information that other researchers can use to link genetic variants to the risk for specific complex illnesses, including cancer, which will lead to new methods of preventing, diagnosing, and treating complex diseases. An additional assumption behind this approach is the idea of the *common variant common disease hypothesis*. If, however, the disease is caused by a rare variant, this approach may fail to detect association. Where haplotype diversity exists, particularly informative SNPs that best characterize a haplotype can be used to limit the amount of laboratory and analytical work in haplotype-based studies. Use of haplotype block information has also been proposed to increase power by 15%–50% when compared to a SNP-based approach [7].

5.2.5 Genome-Wide Association Screen

A genome-wide association screen (GWAS) is an approach that involves rapidly scanning markers (SNPs or haplotype) across the complete sets of DNA, or genomes, of many people to find genetic variations associated with a particular complex disease or cancer [8, 16]. Candidate gene variants, however, may not be studied directly, but gene discovery studies can be accomplished using a strategy that relies on LD between genetic variants. This represents the underlying premise behind whole-genome SNP scans. The whole-genome association approach can identify new candidate genes or regions. Some National Institutes of Health (NIH) institutes have already completed GWASs and deposited their data in the NCBI Database of Genotype and Phenotype (dbGaP) (http://www.ncbi. nlm.nih.gov/ entrez/query.fcgi?db=gap). These studies include research by the National Eye Institute (NEI) on age-related eye diseases and the National Institute of Neurological Disorders and Stroke on Parkinson's disease. Similar successes have been reported using GWASs to identify genetic variations that contribute to the risk of type 2 diabetes, heart disorders, obesity, Crohn's disease, and prostate cancer, as well as genetic variations that influence response to antidepressant medications. Results of genome-wide SNP association studies might not be easily replicated in subsequent studies but could still identify causative regions of the genome. Similarly, large-scale GWASs may find surrogate markers that will distinguish complex chronic disease cases from controls but may not identify causative SNPs. Therefore, replication of associations is crucial to lead to valid and causative associations.

A number of GWASs and candidate-pathway studies suggest that genetic variations contribute to human disease risk, disease phenotypes, and prognosis [8]. For the majority of genetic variants, very large relative risks are needed for a single factor to achieve a high level of predictive value for disease risk. Candidate pathway approaches have generated interesting findings. Nonetheless, it is generally limited by the knowledge and coverage of the underlying etiology. Inherited variations in gene response may be up- or downregulated by chronic environmental exposures and drug exposure, which further complicate the issue.

5.2.6 Epigenetic Changes and Variation in Noncoding DNA Elements

The epigenome consists of the DNA methylation marks and histone modifications involved in controlling gene expression [17]. It is accurately reproduced during mitosis and can be inherited transgenerationally. Imprinted genes and metastable epialleles represent two classes of genes that are particularly susceptible to environmental factors and stressors because their regulation is tightly linked to epigenetic mechanisms [18]. Three genomic targets that are likely to be susceptible to gene expression changes due to environmental influences on epigenetic marks are the (i) promoter regions of some housekeeping genes, (ii) transposable elements that lie adjacent to genes with metastable epialleles, and (iii) regulatory elements of imprinted genes. These genomic targets contain regions that are rich in CpG dinucleotide sequences, which are normally unmethylated, methylated, or differentially methylated, respectively. The methylation status and, in some cases the status of histone modifications in the same region, determines levels of gene expression.

The microarray assay can be applied as a useful tool for mapping methylation changes in multiple CpG loci and for generating epigenetic profiles in various complex chronic diseases. It is important to identify imprinted genes as epigenetically labile genomic targets in the whole genome. The genome-wide identification of the full suite of epigenetically labile targets in human genomes has the potential to link the environmental influence, through epigenetic changes, to early developmental influences on susceptibility to diabetes, obesity, abnormal lipid metabolic syndrome, cardiovascular disease [19], cancer, and behavioral disorders. If environmentally induced epigenetic adaptations occur at crucial life stages, they can potentially change behavior, disease susceptibility, and survival [17, 18]. Epigenetic modifications do not alter gene sequence, but they effect gene expression. Therefore, characterizing the expression profiles of epigenetically labile genes that are susceptible to environmental dysregulation will ultimately identify epigenetic signatures for chronic disease and environmental exposure. These epigenetic signatures can prove to be biomarkers that will allow for early prediction for individuals with a possibility of adult onset of complex disease [18]. They can also be used in novel preventative and therapeutic approaches before disease symptoms develop. Such an approach to human disease management could revolutionize medical care, which now mainly treats diseases only after they develop.

The human genome encodes the blueprint of life, but the function of the vast majority of its nearly 3 billion bases is unknown. Recent advances in creating the Encyclopedia of DNA Elements (ENCODE) project has systematically mapped regions of transcription, transcription factor association, chromatin structure, and histone modification [20]. These enabling data may help us to assign biochemical functions for 80% of the genome, particularly outside of the well-studied protein-coding regions. Many discovered candidate regulatory elements are physically associated with one another and with expressed genes, providing new insights into the mechanisms of gene regulation [5]. Overall, the project provides new insights into the organization and regulation of human genes and the genome and is an expansive resource of functional annotations for biomedical research. The newly identified elements also show a statistical correspondence to sequence variants linked to human disease and can thereby guide interpretation of this variation in the susceptibility or resistance to complex chronic diseases.

5.2.7   Variations, Mutations, and Damages in the Mitochondrial Genome

Mitochondrial DNA (mtDNA) is almost exclusively inherited from the maternal line. Stable SNPs have emerged in mtDNA over the past 150,000 years [4]. The distribution of these mtDNA polymorphisms varies greatly across populations, reflecting human migration and adaptation to environmental conditions and exposures to stochastic factors. The result is that in humans there are different types of mitochondrial electron transport chains with different capacities for energy production and reactive oxygen species (ROS) generation. There is growing evidence that certain mtDNA clusters are associated with distinct disorders [21–23]. The D-loop is a hot spot for mtDNA mutations, and it contains two hypervariable regions (HV1 at positions 16024–16383 and HV2 at positions 57–372) [24]. Variations in mitochondrial SNPs have been associated with hormone-dependent cancers [22, 23]. The oxidative mtDNA damage per mtDNA copy number is higher in many chronic disease tissues. Recent studies show that specific mitochondrial haplotypes and mtDNA SNPs variations can contribute to human disease risk [21]. Mitochondrial lineage haplogroups can be used to estimate ancestral proportions in the populations manifesting various complex chronic illnesses. To investigate whether variations in common alleles or haplogroups of mtDNA are independently associated with mtDNA damages/mutations, potential environmental and stochastic factors, disease subgroups, and covariates (sociodemographic information, alcohol/drug use, smoking, body mass index, and diet) should be included in a GEI testing model.

5.2.8   Chronic Disease Bionetwork through Interactome, Phenome, and Systems Approaches

Traditional epidemiological and biostatistical association study approaches have limitations in the prediction of actual biological causality of the progression of chronic diseases. Evidence of candidate gene or SNP variants association is not a proof of the increased or decreased level of protein expressions. In addition, these associations also do not indicate that how the environmental factors may influence protein–protein interaction to augment or slow down the progression of complex diseases. Such studies also lack the power to identify the posttranslational modifications of expressed proteins due to epigenetic hypermethylation of regulatory elements of genes harboring CpG elements or through alternate mechanism. To fully assess GEI, a more comprehensive understanding of the physiological, cellular, and molecular effects is required within the context of the whole organism and its transcriptome, proteome, and metabolome [4]. It is important, therefore, to capture protein interaction networks in humans to correlate with biological complexity in the manifestation of complex diseases. In the recent past a massive volume and variety of "-omics" big data have been generated and curated. It is important that study design incorporate data from many "-omics" sources to predict biological plausibility that may provide a real-time picture of complex cellular processes that may be directly involved in the progression and development of complex diseases. Only then studies can develop compatible intervention plans at the diagnostic or therapeutic level. There are a number of data-mining tools already available that have been developed by researchers, or very often, they are government supported. The following is a list of some of the tools and related databases that can be used to design studies to determine the biological causality of the complex diseases. The flow of the "-omics" tools after genomic association studies (GWASs, SNPs, etc.) may include information from transcriptomics, proteomics, interactomics, and metabolomics technologies in individualized therapy and diagnosis for chronic diseases. These are also being continuously updated.

5.2.8.1 Transcriptomics

Transcriptome is the dynamic population of functional RNA transcripts that varies on the basis of the time, cell type, genotype, external stimuli, and mechanisms that regulate the production of RNA transcripts. Therefore transcriptomics include characteristics and regulation of the functional RNA transcript population of a cell or organism at a specific time. The sequence of an RNA mirrors the sequence of the DNA from which it was transcribed. Consequently, by analyzing the entire collection of RNAs (transcriptome) in a cell, investigators can determine when and where each gene is turned on or off in the cells during the progression and development of complex chronic diseases. The National Human Genome Research Institute (NHGRI), which is part of the NIH, has participated in two projects that created transcriptome resources for use by researchers around the world. Those projects were the Mammalian Gene Collection Initiative and the Mouse Transcriptome Project.

5.2.8.2 Proteomics

Proteomic analysis involves protein mapping or profiling to identify the property of the primary amino acid sequence. Quantitation of all proteins from a defined space is inherent in protein profiling, and isolation or enrichment of proteins from a particular spatial location within cells or tissues helps to characterize the organism's phenotype. Proteomic platforms represent strategies for global separation and identification of proteins. Separations are generally accomplished by gel electrophoresis, although more recent studies incorporate liquid chromatography (LC)-based platforms, such as linear column gradients or multidimensional chromatography (MuDPIT). Mass spectrometry (MS) is the primary means of protein identification in proteomic analysis. Identification occurs by peptide mapping or amino acid sequencing. Retentate chromatography MS has been used for rapid profiling of biofluid samples using chemically reactive surfaces for separation and matrix-assisted laser desorption/ionization (MALDI) for generating protein mass spectra (i.e., surface-enhanced laser desorption/ionization (SELDI) technology). However, alternatives to MS-based identification in proteomic analysis exist in platforms based upon affinity arrays such as antibody arrays, antibody multiplexing, and fluorescently tagged antibodies bound to bead suspensions such as Luminex technology.

5.2.8.3 Interactomics

When the network of complex protein–protein interactions (PPIs) in a cell or organism is reconstructed, the result is called an interactome. This complete network of PPIs is now thought to form the backbone of the signaling pathways, metabolic pathways, and cellular processes that are required for all key cell functions (cell differentiation, apoptosis, cell survival, etc.) and in the origin and progression of complex chronic diseases. The interactome allows high-throughput studies to investigate systematically many potential interactions in a variety of models, which provides unique opportunities to compare interaction networks and ask questions about their conservation during complex diseases. It is expected that this approach will bring a closer resolution on identifying candidate pathways leading to a complex disease in an individual. Such a complete knowledge of cellular pathways and processes in the cell is essential for understanding how many complex diseases originate and progress through mutation or alteration of individual pathway components. Furthermore, determining human cellular interactomes of therapy-resistant tumors will undoubtedly allow for rational clinical trials and save patients' lives through designing a tailored therapy for an individual.

5.2.8.4   Metabolomics

Metabolites, the chemical entities that are transformed during metabolism, provide a functional readout of cellular biochemistry. Metabolomics, an "-omic" science in systems biology, is the global quantitative assessment of endogenous metabolites within a biological system. Either individually or grouped as a metabolomic profile, detection of metabolites is carried out in cells, tissues, or biofluids by either nuclear magnetic resonance spectroscopy or MS. With emerging technologies in MS, thousands of metabolites can now be quantitatively measured from minimal amounts of biological material, which has thereby enabled systems-level analyses. By performing global metabolite profiling [25], also known as untargeted metabolomics, new discoveries linking complex cellular pathways to biological mechanisms are shaping our understanding of the origin and progression of complex chronic diseases [26].

5.2.8.5   Phenomics

Complex diseases represent a clustering of many phenotypes in response to the exposure of many environmental and stochastic factors though the life stages. For example, in a metabolic syndrome case chronically elevated blood pressure, cholesterol levels, and plasma glucose, as well as abdominal obesity, are associated with an increased risk of atherosclerosis and type 2 diabetes. Current approaches to identifying unifying genetic mechanisms (i.e., pleiotropy) remain largely focused on clinical categories that do not provide adequate etiological information [27]. Therefore, as an alternative, the phenomics approach that assembles coherent sets of phenotypic features that extend across individual measurements and diagnostic boundaries for novel genetic investigations of established biological pathways and complements the traditional GWAS study or candidate gene-pathway-based strategy focused on individual phenotypes [27]. Phenomics is the use of large-scale approaches to study how genetic instructions from a single gene or the whole genome translate into the full set of phenotypic traits of an organism.

5.3 Integration of Functional Genomic, Epigenetic, and Environmental Data of Molecular Epidemiological Studies Using Bioinformatics and Biostatistics Approaches

Several steps will be required in the integration of genomic data on mutations, SNPs, CNVs, molecular interactions, epigenetic and transcriptomic changes that influence gene expression, the proteome, and the metabolome that trigger the onset and progression of or resistance to complex chronic diseases. Further, integration of bioinformatics, statistical, and validation approaches will need to be employed. These steps can broadly include (i) molecular epidemiologic study designs for G × G and GEI, (ii) identification of genomic data on mutations, SNPs, changes in CNVs, and variations in a significant set of over- and underexpressed genes, (iii) enrichment analysis of a set of significant genes and proteins, (iv) analysis of combined effects of modified gene expressions, CNVs, mutations, and molecular interactions influencing biological pathways and network(s) that contribute to the susceptibility to, resistance to, or development of complex diseases, (v) validation of key causal genes/proteins/molecules/ environmental and stochastic factors predicted to be involved in the development of disease by biostatistical approaches, and (vi) literature-based validation of key causal genes/proteins/molecules/environmental and stochastic factors involved in the etiology of chronic diseases using models generated by empirical data. A brief description of these steps follows.

5.3.1 Molecular Epidemiologic Study Designs for G × G and GEI

Epidemiological studies have been remarkably successful in identifying the main risk factors for many common diseases due to the best available study designs and data-collection methods. The relative merits of population-based epidemiological studies are well established. Studies in genetic epidemiology, however, have been dominated by the use of family-based designs from which inherited susceptibility can be inferred. The common approaches to design studies for G × G and GEI include family-based studies, studies of unrelated individuals, retrospective design, prospective design, and case-only design.

5.3.1.1 Family-Based Studies

By comparing disease concordance rates between monozygotic and dizygotic twins, twin studies can be used to partition components of variance between genetic and shared and nonshared environmental factors. Most reports of studies from twin registries do not include information on environmental exposures that could be shared (or different) between the twins, precluding any inferences about specific GEI. Analyses of multigenerational pedigrees might provide a preliminary assessment of the hypothesis whether penetrance of a mutation has changed over time, which would indicate that changes in lifestyle and environment influence gene penetrance [4, 28]. A limitation of this approach is that assessments of this nature can only be made for relatively highly penetrant gene mutations (i.e., where the penetrance is sufficiently high to cause clear familial aggregation). Incorporation of environmental data into pedigree or other family-based designs (e.g., studies that use sib-pairs or case–parent designs) allows direct estimates of specific GEI. Collection of adequate numbers of sib-pairs, however, may require more effort than using unrelated controls and, for the late onset of chronic diseases and cancer, availability of living parents may limit case–parent accrual [28].

5.3.1.2 Studies of Unrelated Individuals

With the advent of methods for assessing DNA sequence variability directly, GWASs using unrelated individuals are increasingly being used. It should also be noted that, under certain assumptions, GEI can be estimated from case–case studies without controls. However, the search for GEI imposes constraints on the use of these designs. In retrospective case–control studies, data on environmental and lifestyle factors and samples for DNA and biomarker studies are obtained after the diagnosis of chronic disease in the cases. In prospective cohort studies, environmental and lifestyle data are obtained at the baseline (the start of the study) and ideally at other points before diagnosis. Samples for DNA and biomarker studies are also ideally obtained at baseline, although in prospective studies that do not have banked samples, DNA can be obtained after diagnosis from living cohort members [28].

5.3.1.3 Retrospective Design

The chief advantage of case–control retrospective studies is the potential for the sample size to be limited only by cost and by the number of cases of the disease that are available in the study area [28]. It is always a concern, however, that in retrospective studies, poor recall (misclassification) of past exposures among both cases and controls might attenuate the estimates of risk to the point where any difference in risk according to genotype cannot be reliably detected for GEI. The primary limitation of retrospective studies, particularly in case–control studies, is the selection bias (in particular, the use of controls that do not represent the population in which the cases occurred) and confounding [28]. With respect to GEI, the misclassified ("noisy") or biased information on environmental exposures is a major challenge. Bias can arise if cases report their prediagnosis exposure histories differently once they are diagnosed with a disease compared to what they would have reported before diagnosis (recall bias). These biases and misclassification can be reduced, but rarely eliminated, by paying careful attention to the best practices in enrollment and exposure assessment. If the race or ethnicity of the controls is substantially different from that of the cases, then spurious associations with gene variants that differ by race or ethnicity (i.e., population stratification) will occur. While designing retrospective case–control study, the potential influence of population stratification can be substantially eliminated with attention to appropriate choice of controls and by controlling for self-reported ethnicity. Methods to assess the population substructure of cases and controls by genotyping noncausal gene variants (genomic control) have been proposed in retrospective studies and can be used to correct for this phenomenon. Given the need for large sample sizes in gene–environment studies, retrospective studies can be the method of choice with a potential advantage only when they are combined with increased scrutiny of exposure assessments for controls and cases for the estimation of chronic and complex diseases and some type of cancer phenotypes.

5.3.1.4 Prospective Design

The main advantage in the design of prospective studies is that DNA samples and exposure information are obtained from participants in a longitudinal cohort who are followed up usually for years or decades. In this design, an attempt can be made to obtain DNA samples from cases arising in the cohort and from matched or unmatched noncases. The problems of selection and recall bias in case–control studies can also be minimized in prospective studies. If follow-up rates are high, then a virtually complete set of cases can be assembled and compared with a sample of individuals who did not develop the disease. The use of this nested case–control study minimizes selection bias because the population that gave rise to the cases can be defined [28]. Because information on exposures is collected before diagnosis (in most cases, years to decades before), recall bias is eliminated as knowledge of diagnosis cannot influence the reporting of exposures. In many cohort studies that only have a baseline assessment and do not involve repeated measurements during follow-up, a single measure of an exposure might not be a good reflection of the pattern of exposures on various life stages. The concern of differential participation by cases and controls according to ethnicity that potentially gives rise to population stratification (especially in populations whose ancestors have been recently mixed by intercontinental migration) can be minimized by collecting information on controlling for ethnic background or by using genomic tools in the design of prospective studies. The principal problem with prospective studies is that adequate sample sizes of cases can only be obtained for common conditions, such as hypertension, myocardial infarction and stroke, and common cancers, in the population that is being followed [28]. Rare diseases, such as sarcomas, will not occur at sufficient frequency to provide statistical power. A typical cohort study might only accrue several hundred cases of a disease of moderate incidence (e.g., Parkinson's disease) over many years, and because most cohort studies enroll men and women in middle life, diseases with relatively early onset (e.g., multiple sclerosis) will be underrepresented. In addition, some of the special requirements for genomic analyses might only be met in case–control studies. Obtaining the fresh-frozen tissue or tumor blocks necessary to subtype these outcomes might be difficult in prospective studies but more feasible in cases that are studied in a limited number of institutions. Chronic and complex diseases are clusters of many diseases, and samples collected manifest different stages of progression. In prospective studies that include "-omic" tools, expression-array analyses may be able to delineate various stages of a disease that may look histologically similar. Phenotypic assays, such as assays that measure the activity of an enzyme or a biochemical pathway by giving a test dose of a compound and measuring metabolites in blood, might be possible in a limited number of cases and controls but are unlikely to be feasible in large prospective studies due to logistical challenges, such as duration of the executed study and institutional review board (IRB) approval. Information on disease diagnosis and subtype from nongenomic tests, such as histology or imaging, might also be hard to obtain in a uniform manner in a prospective study, in which almost all cases might be diagnosed at different institutions, as opposed to a case–control study that operates in a limited number of facilities [4, 28].

5.3.1.5 Case-Only Design

It has been shown that when a genotype is not correlated with environmental and stochastic factor(s) and a disease is rare (few controls are likely to have an undiagnosed or incipient disease), then departure from multiplicative interaction can be tested by examining information from the cases only. In this case–case design, the prevalence of the exposure in the genotype-positive cases would be expected to be the same as the prevalence of the exposure in the genotype-negative cases [28]. Statistically significant departures from this expectation of equal prevalence indicate an interaction between genotype and exposure. To estimate these, an appropriate control group is needed. It is possible that, as we identify the main genetic influences on chronic diseases and cancers, case–case methods might become more popular in assessing the interaction of established causal genotypes with environmental and stochastic factors, particularly as these studies can form the baseline cohort for finding prognostic markers of disease outcomes [4].

5.3.2 Identification of Mutations, SNPs, Changes in CNVs, and Variations in a Significant Set of Over- and Underexpressed Genes

As described earlier with the completion of the Human Genome Project in 2003 and the International HapMap Project in 2005, researchers have a set of research tools to find the genetic contributions to complex diseases. As studies of genetic susceptibility and environmental exposures have been largely pursued by different groups of investigators, multidisciplinary collaboration is necessary to generate the best studies in the field. Obtaining high-quality data on the environment and lifestyle, in conjunction with biological samples to assess these genetic variants, is crucial in the assessment and integration of GEI.

In addition to obtaining data from prospective studies, integration of toxicological and pharmacological public databases such as the Comparative Toxicogenomics Database (CTD) and the Environmental Genome Project (EGP) with data on genetic variation has proven useful in comprehensive understanding for research into GEI-related complex diseases. A flowchart of the data integration is shown in Fig. 5.2. While following a similar flowchart Kunkle, Yoo, and Roy demonstrated the use of the comprehensive bioinformatics method to map GEI in the glioblastoma [3]. The study used genetic variations (due to CNVs, SNPs/insertions/deletions, or mutations) and environmental data integration that links with specific cancer to identify (i) genes that interact with environmental pollutants and chemicals and have genetic variants linked to the development of cancer, (ii) important pathways that may be influenced by environmental exposures (or endogenous chemicals), and (iii) genes with variants that may have been understudied in relation to carcinogenesis. Findings in similar studies that use a comprehensive integrative approach will reveal the potential role of GEI in the onset and progression of complex diseases.

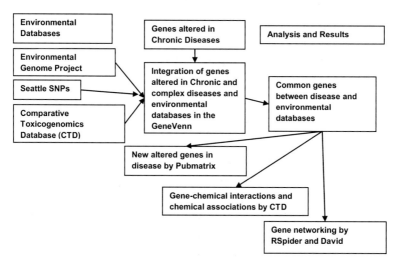

Figure 5.2   Flowchart of the bioinformatics method showing genetic alterations (copy number, mutations, and SNPs) and environmental data integration that links with a disease to identify genes that interact with chemicals and have linkage with a particular chronic disease.

Genes of NIH-sponsored public environmental databases can be used for the cross-referencing of environmentally important genes with genes with variants associated with complex diseases. These databases focus on the validation of the environmentally responsive genes included in them, either through laboratory work in the databases projects themselves or through literature. For example, some of the databases are:

- The EGP, located at http://www.niehs.nih.gov/research/supported/programs/egp/: The National Institute of Environmental Health Sciences (NIEHS) EGP is a database that identifies and genotypes genes with functions related to cell cycle, cell division, cell signaling, cell structure, DNA repair, gene expression, homeostasis, metabolism, immune and inflammatory response, hormone metabolism, nutrition, oxidative metabolism and stress, membrane pumps and/or drug resistance, and signal transduction. According to Kunkle, Yoo, and Roy (2013), all finished genes from this database were included for comparison with the genes known to contain variations in glioblastomas [3]. The list of finished genes, which are those genes found to be responsive to environmental factors by EGP researchers, can be found at http://egp.gs.washington.edu/ finished_genes.html.

- Seattle SNPs, located at http://pga.gs.washington.edu/: The Seattle SNP database is a National Heart Lung and Blood Institute (NHLBI)-funded project focused on identifying, genotyping, and modeling associations between SNPs in candidate genes and pathways that underlie inflammatory responses in humans. The list of finished genes, which are those found to be responsive to environmental factors by Seattle SNP researchers, can be found at http://pga. gs.washington.edu/finished_genes.html.

- The CTD, located at http://www.mdibl.org/research/ ctd.shtml: The CTD, operated by the Mount Desert Island Biological Laboratory (MDIBL) with support from the NIH, collects gene–environment–disease interactions information from published literature. It is searchable by several methods, including disease, gene, chemicals, and gene–chemical interactions. A chronic disease can be searched to obtain a list of genes found to be modulated by chemical exposures in the specific disease. The CTD selects genes for the chemical-associations-to-disease search if they have either (i) a curated association to a disease through a marker/mechanism or therapy or (ii) an inferred association via a curated chemical association.

- The variant genes from the curation and environmental databases can then be inputted into the MyGeneVenn program to assess their overlap, as depicted in Fig. 5.3, while using a hypothetical scenario. Common genes between the three environmental gene databases and specific variant genes can be selected. Both these overlapping (common) sets of genes can be referred to as environmentally responsive interacting genes (GEI genes) from this point forward, and the entire list of variant genes can be used for further analysis.

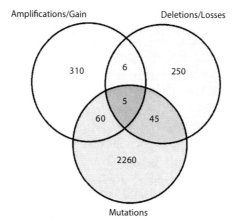

Amplifications/Gain   Deletions/Losses

Mutations

Figure 5.3   A typical MyGeneVenn analysis of an overlap between chronic disease-specific amplified/gain genes, deleted/loss genes, and mutated genes.

5.3.3   Enrichment Analysis of Significant Genes and Proteins

The final set of altered genes, SNPs, CNVs, mutations, and variations in epigenomes, proteomes, transcriptomes, and metabolomes identified in the molecular epidemiologic studies can be subjected to enriched pathway analysis with several different tools. FuncAssociate (Roth Laboratory, Harvard) and Ingenuity Pathway Analysis (IPA) (Redwood City, California) are examples of web-based tools that can be used to identify pathways leading to complex chronic diseases. These tools can also help in integrating characteristic biological pathways of complex diseases that may be due to altered gene functions.

5.3.4   Analysis of Combined Effects of Modified Gene Expressions, CNVs, Mutations, and Molecular Interactions Influencing Biological Pathways and Network(s) That Contribute to the Susceptibility to, Resistance to, or Development of Complex Diseases

It is difficult to extract relevant information on G × G and GEI from population-based association studies that contribute to the susceptibility to, resistance to, or development of complex diseases. Stochastic modeling based on the mechanistic understanding is critical for a successful integration of G × G and GEI. To directly investigate interactions among genes, an analysis of effects of all genes together is essential. Though conventional statistical regression methods may be able to analyze effects of all genes together, when the number of possible G × G grows exponentially in each stage of a chronic disease, then the number of interactions to be estimated becomes too large for conventional statistical regression methods. The inability of conventional statistical regression to compute high-order complex interactions has encouraged the recent use of model-free methods, such as multifactor dimensionality reduction (MDR), and graphical means for representing causal relations, such as structural equation models (SEMs) and causal Bayesian networks (CBNs). We will further discuss MDR and CBN.

5.3.4.1 Multifactor Dimensionality Reduction

MDR is a method that reduces the number of modeled variables by combining variables into a single variable. This approach overcomes the typical problem of genomic data that are sparse, that is, a large number (>100) of modeled variables (genes, environmental factors, etc.) and a small number (<30) of measured cases, that is referred to as the curse of dimensionality [29]. MDR analysis has been widely used for G × G analysis. In practice, however, it is not easy to perform high-order G × G with ever-growing genomic data, for example, when more than 1000 genes are to be analyzed via MDR at the genome-wide level, because it requires exploring a huge search space and suffers from a computational burden due to high dimensionality [30].

5.3.4.2 Bayesian Networks

Bayesian networks (BNs), also known as probabilistic causal networks, is one such approach that is increasingly being used in the modeling of uncertain knowledge. A Bayesian model search is flexible, robust, and computationally efficient and lends itself naturally to the creation of genetic risk classifiers [31]. Bayesian classifiers have been used before in GWASs but generally only on individual phenotypes. Bayesian classifiers produce classification rules that are equivalent to using logistic regression with a genetic risk score. The advantage of the model approach based on Bayesian classifiers is that it can be generalized to include multiple traits, in G × G or GEI models. Even pleiotropic associations can be directly modeled via the construction of a simple BN. The BN also overcomes the shortcomings of the simple Bayesian method, first explored extensively in the early 1960s. On a given hypothesis, a BN assumes conditional independence of the evidence that can model causality on the basis of an expert's knowledge, data, or both. In a causal BN (or a causal network for short), each arc is interpreted as a direct causal influence between a parent node (variable) and a child node relative to the other nodes in the networks. The BN has been used in many clinical studies because of its ability to find inference in the forward direction, backward direction, or any combination. Similar to any powerful reasoning methodologies, one practical limitation of BNs is that inference within them is NP-hard. Researchers have developed approximate inference methods (to address the NP-hard inference issue) to apply to complex BN inference tasks. Some of the software programs and tools that are currently available are as described below.

5.3.4.2.1 *Banjo*

To further investigate the complex G × G and GEI to understand the etiology of chronic disease, a software application similar to Banjo (developed at Duke University, North Carolina) can be used for probabilistic structure learning of a static BN. To perform the analysis on differentially expressed genes, SNPs, changes in gene copy, mutations, and variations in epigenomes, proteomes, transcriptomic and metabolomic data, and values for a significant set of molecules can be uploaded into Microsoft Excel for analysis in Banjo. Banjo performs structure inference using a local search strategy termed the "Bayesian Dirichlet equivalence" (BDe) scoring metric for discrete variables. This strategy makes incremental changes in the structure aimed at improving the score of the structure. A score for the best network, influence scores for the edges of the best network, and a dot graphical layout file are returned as a result of the search. The dot file is a directed acyclic graph (DAG) indicating regulation among genes and their possible influence on chronic disease outcome.

Banjo has been shown to have a very high positive predictive value for 100-plus case sets (regardless of the number of genes) composed of the type of global, steady-state gene data.

5.3.4.2.2   *Markov blanket*

The BN can be used to identify what may be the most critical genes for development of chronic and complex diseases. This can be accomplished by identifying a Markov blanket of each network output chosen as the best network for each grade of disease. In a BN, the Markov blanket of any node A is its set of neighboring nodes composed of a nodes parents, children, and the parents of the children. This defined set of neighboring nodes shields node A from the rest of the network, and thus the Markov blanket of node A is the only knowledge needed to predict the behavior of node A [32]. An example of Markov blanket genes for a BN of anaplastic astrocytoma is depicted in Fig. 5.4.

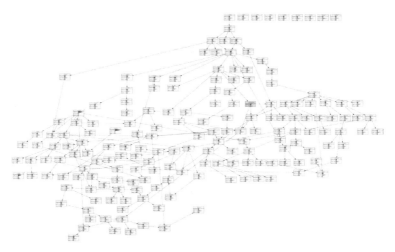

Figure 5.4   Example of Markov blanket genes for the Bayesian network of anaplastic astrocytoma. Green-shaded genes are overexpressed genes, and blue-shaded genes are underexpressed genes from the oncomine meta-analysis.

5.3.4.2.3   *Perl*

To increase our sample size, imputation of missing cases can be performed on cases with missing expression data for a particular molecule using the average of all values across the molecules. A program Perl can be used to analyze the data multiple times separately (three hours in length for each network search) to select the final best network for each category representing different sample cohort, (e.g., controls, different stages of cancer). The best network score significance is calculated using a log calculation of all three network scores, with a percentage of the total score returned for each network [2].

5.3.5 Validation of Key Causal Genes/Proteins/ Molecules/Environmental and Stochastic Factors Predicted to Be Involved in the Development of Disease by Statistical and Other Approaches

Several methods can be used to validate both the prediction capabilities and the biological plausibility of selected Markov network genes/proteins/molecules/environmental and stochastic factors. It is essential that studies include one or more of the following approaches to validate the findings from the BN or Markov blanket analysis to further enhance the power to predict the susceptibility and/or resistance to complex diseases.

5.3.5.1 Validation by Statistical Methods

Several statistical methods can be used to both validate the prediction capabilities and assess the ability of Markov genes/proteins/molecules/environmental and stochastic factors to distinguish between normal and disease samples. However, a prediction analysis by a receiver operating characteristic (ROC) curve representing the BN results can be used to validate the findings using linear regression, logistic regression, cross-validation, and support vector machine (SVM) analysis to assess the predictability of both the discretized and raw expression values of the Markov genes/proteins/molecules/environmental and stochastic factors. These kinds of investigations, though, produce an array of statistical results, but the validity of these results to discern more about the biological basis of obtained associations with complex diseases and nondisease can also be done by hierarchical clustering of a set of Markov genes/proteins/molecules/environmental and stochastic factors [33]. These analyses can be performed by using standard IBM SPSS Statistics 19.0 and/or Multi-Experiment Viewer (MeV) version 4.7.1 or similar computer software.

To validate selected Markov genes/proteins/molecules/ environmental and stochastic factors, structural equation modeling (SEM) can also be used. SEM has an advantage in validating causal relationships between genes or environmental factors and genes. Such integrative models may present a way to account for hidden individual differences in susceptibility to diseases. The capability of SEM in empirical validation, combined with the prediction and diagnosis capabilities of Bayesian modeling, should facilitate the recognition of genes related to disease outcome, from identification of causal relationships to understanding the mechanisms of disease, designing drugs, and treating patients [34].

5.3.5.2 Literature-Based Validation of Key Causal Genes/ Proteins/Molecules/Environmental and Stochastic Factors Involved in the Etiology of Chronic Diseases Using Models Generated by Empirical Data

This includes biological database searches, literature reviews, curated genes, and pathway analysis. The literature and database search of Markov genes gathers information on cellular localization of gene(s) and their function and published research supporting the genes' involvement in complex disease initiation and development, for example, the Human Gene Compendium's Gene Cards (www. genecards.org), PubMed (www.pubmed.com), the Information Hyperlinked over Proteins (iHOP) Database (www.ihop-net.org), and the Individual Chronic and Complex Disease Databases.

To investigate existing literature- and ontology-based connections between selected Markov gene/proteins/molecules/ environmental and stochastic factors, programs like IPA and PathJam can be used. The goal of literature-based validation is to ascertain (i) the quality of network analysis findings in programs like Banjo and GeNIe and (ii) the biological relationships of selected Markov genes/ proteins/molecules/environmental and stochastic factors from these analyses. Web-based tools like IPA and STRING have curated literature and experimental evidence of biochemical interactions to produce networks of existing connections between a set of user-inputted genes and proteins. The Path Explorer and Core Analysis features of IPA also allows initial validation to search for any existing connections and then to produce connections in the complex disease and Markov key causal genes/proteins/molecules/environmental and stochastic factors, which can even be independent of their established pathways. Networks are organized by importance by an IPA-defined significance score. Complementary to the IPA's Core Analysis production of top biological and disease-related functions can be carried out using PathJam [35]. This public server–based tool allows for interpretation of gene lists by integrating pathway-related annotations from several public sources, including Reactome, KEGG, NCBI Pathway Interaction Database, and Biocarta. Using this tool, interactive graphs can be produced, linking selected Markov gene lists with pathway annotations, allowing for graphical pathway investigation into gene lists. A graphical depiction of similar analysis in the case of astrocytoma is shown in Fig. 5.4.

5.3.5.2.1 *Text-mining search*

Pubmatrix (http://pubmatrix.grc.nia.nih.gov) is an NIH tool that allows cross-referencing of gene lists with search terms to assess whether genes interacting with environmental and stochastic factors had been previously reported as significant to specific complex diseases. The Pubmatrix search uses an algorithm to match user-inputted lists to gene names/symbols, etc., from abstracts, keywords, and titles of studies in Medline. Thus the number of mentions a gene receives in these locations can be taken as a proxy for their relative importance to date for research focus in relation to a complex disease or carcinogenesis.

5.3.5.2.2 *CTD search*

The genes interacting with environmental pollutants and chemicals can be searched in the CTD for relevant association with a given disease. For example, a search for "chemical or environmental pollutant associations" produces a list of chemicals that influence a gene to cause initiation or progression of a disease. Likewise, a search for "gene–chemical interactions" produces a list of actions caused by a chemical on a gene function, such as promotion of mutagenesis or increase of expression.

5.3.5.2.3 *Gene networking analyses*

Both the GEI gene list and the complex disease variant gene list can be subjected to gene networking analysis using the bioinformatics tools RSpider and DAVID. RSpider results are visualized using Cytoscape [3].

5.4 Predictive Analysis of Lifetime Risk of Developing Disease

Software like GeNIe, developed by the University of Pittsburgh, can be used to analyze BNs to assess the predictive ability of Markov genes/proteins/molecules to distinguish between normal and disease samples [36]. Once network parameters are established in GeNIe and data are uploaded using Microsoft Excel, the probability of developing each grade or stage of chronic disease can be calculated by using Bayes's rule [2]. GeNIe allows for predicting the probability of developing a disease due to certain expression states of the molecules in their respective networks. This approach may identify key Markov causal genes/proteins/molecules/environmental and stochastic factors involved in the development and progression of chronic diseases. In BN analysis identification of key Markov causal genes/proteins/molecules/environmental and stochastic factors is achieved by learning the parameters of a given DAG structure.

There is a general formula as shown in Fig. 5.5 to further examine collaborative and interactive effects of Markov blanket genes on the lifetime risk of developing cancer or chronic disease. It is possible to use the surveillance, epidemiology and end results (SEER)-calculated lifetime probability of diagnosis of cancer and/or chronic disease [37].

$$P(D|G_1,G_2,...,G_n) = \frac{P(G_1,G_2,...,G_n|D)P(D)}{P(G_1,G_2,...,G_n)}$$

Figure 5.5 General formula to calculate the lifetime risk of developing cancer or a chronic disease, where G_1, G_2, . . ., G_n are selected Markov blanket genes/proteins/molecules/environmental and stochastic factors from the BN with the highest BDe score and D represents whether a subject has cancer or chronic disease.

5.5   Technical Challenges

The flowchart shown in Fig. 5.6 presents a daunting task of data integration from a multitude of genetic, epigenetics, bioinformatics, environmental, and epidemiological platforms. It is essential that future research be directed to address the following technical challenges hindering advancement in the understanding of the mechanistic basis of GEI leading to the origin and progression of complex chronic diseases.

5.5.1   Sample Size

The need for large sample sizes in a GEI study design to integrate genetic, environmental, and stochastic factors relevant in complex human diseases will remain a major challenge. If information on exposures is misclassified, then the power to detect interactions is attenuated, and even larger sample sizes will be required. As such, epidemiological studies are often underpowered for main effects that will be inadequate to detect interactions. Even then, to convincingly show the main effect of a single factor might require a meta-analysis of many studies, and it is rare for the level of detail to be available to identify interactions. A potentially effective means of mitigating the lack of power in prospective studies is to pool data across these studies for meta-analysis [38]. Maximizing the power of ongoing prospective studies in this manner can mitigate the main weakness of prospective studies (i.e., the limited number of incident cases), while capitalizing on the methodological strengths of the prospective design [28]. The meta-analysis of GEI studies may have at least one advantage over conventional two-way environmental interactions because it should be possible to measure a defined functional genetic polymorphism that has an influence on GEI almost without error. However, when several polymorphisms in a gene contribute to altered function, measuring a subset will result in misclassification and so will increase the sample size that is required to detect interactions. Furthermore, functional gene variants are not known—for example, if we are trying to detect genetic association through LD, it is likely that there will be substantial misclassification of the genetic variables, leading to dilution of the relative risk for the interaction. Therefore, in the foreseeable future the occurrence of false-negative findings for interactions in individual studies will be the major problem, unless the interactions are strong. It has been pointed out that most cohort studies will not accrue sufficient number of cases for many complex diseases and might have only marginal power to associate with any meaningful GEI. Therefore, rigorously designed case–control studies will remain the only option for assessing GEI in the majority of complex diseases.

5.5.2 Complex Mixture of Covariates

Which components of complex mixtures—such as air pollution, diet, or cigarette smoke—can cause disease is the most critical question to ask in any analysis of GEI. This is difficult to study observationally as most components of complex mixtures are highly correlated and so their effects cannot be statistically separated. If the effect of the environmental factor differs according to variation in one or more specific genes, then the function of the gene might help to isolate the causal components in the complex mixture [28]. To directly investigate interactions among genes, an analysis of effects of all genes together is essential. Though logistic regression may be able to analyze effects of all genes together, when the number of possible G × G grows exponentially in each stage of a chronic disease, the number of interactions to be estimated becomes too large for logistic regression model. Most of the recent studies on the interactions between genes are often focused on evaluating pairwise interactions. The related concept of Mendelian randomization has been used to argue that a reproducible effect on disease risk of a genotype that alters the level of intermediate G × G is unlikely to be confounded by other lifestyle variables because in most cases, these other lifestyle variables would not be expected to correlate with genetic variation [4].

5.5.3 Coordination in Data Collection and Their Meta-Analysis

Despite much information on both genetic and environmental disease risk factors, there are relatively few examples of reproducible GEI [28]. There are two approaches to mitigate this problem, to facilitate web-based presentations of unpublished results in supplementary tables and to preplan analyses across many studies so that the data are analyzed and displayed in as uniform a format as possible.

The latter approach is a prospective variant of the meta-analysis-of-individual-participant-data approach. The extra advantage of this approach is that preplanned analyses allow more consistent treatment of LD and haplotype definition. With the increasing use of haplotype-tagging or LD-tagging SNPs to explore genetic associations in candidate genes and regions, there is even more potential for incompatible information. Some degree of coordination of the main studies in each disease area would at least reduce the potential for incompatibility of information and could expedite the confirmation of replication of both genetic effects and GEI [28]. The NCI Breast and Prostate Cancer and Hormone-Related Cohort Consortium, for example, is a planned assessment of the same genetic variants in 53 candidate genes across 10 studies that collectively contribute more than 6000 cases of breast cancer and 8000 cases of prostate cancer [14]. Consortia such as this have the potential to provide much more uniform data and analyses than are available through post hoc or literature analyses.

5.5.4 Lack of Computational Power

Massive data on GEI are also creating this big data glut. This huge amount of GEI data has a lot of noise, with potential GEI association inside this noise. Investigators currently face this enormous challenge to delineate between the noise and the correct association between the set of genes and environmental drivers of complex chronic diseases. There is a lack of computational and bioinformatics methods that can reduce large and diverse environmental, epigenetic, epidemiological, and "-omics" data sets into representations that can be interpreted in a biological context. In response to environmental and stochastic drivers, the temporal and spatial regulations of genes at transcriptome, proteome, and metabolome levels indeed require data standards and metadata descriptions of the experiments in a format that enables computational approaches to data analysis. There is a need to establish a computing ability to pull out patterns of signals from these overflowing data streams at all levels that can diagnose validated aberrations causing initiation and progression of complex diseases. More importantly, diagnosis has to happen early and in real time. Late detection of any complex disease or advanced stages of cancer will do little or nothing to prevent poor prognosis. Complex diseases occur, and patients suffer or die in response to environmental and stochastic factors. The challenge is to decipher the biological complexity and prevent a poor prognosis. Therefore, in addition to building a robust computing ability more data from a diverse platform need to be integrated (Fig. 5.6) to develop efficient and validated early diagnosis markers and therapeutic targets of chronic and complex diseases. Then only is it possible to develop and implement a meaningful prevention and treatment plan.

5.6 Conclusion

Figure 5.6 shows an integration model of genomics, transcriptomics, interactomics, or metabolomics to track biological networks holistically, along with hypothesis-driven approaches.

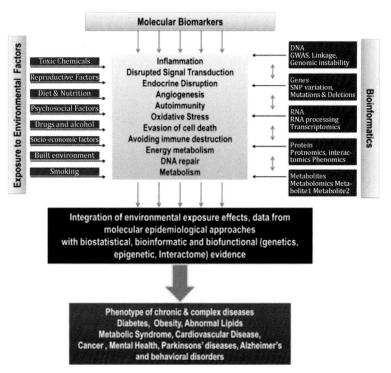

Figure 5.6   Integration model of new discovery-driven technologies, for example, transcriptomics, interactomics, or metabolomics, to track biological networks holistically, along with classic hypothesis-driven approaches.

Different approaches in genetic epidemiology, bioinformatics, and biostatistics to elucidate the structure of biological networks, as well as conceptual and detailed modeling to illuminate how network structure relates to function, play an increasingly critical role in this effort to delineate the pathways of complex chronic diseases. The clear benefit of the integrative biostatistics, bioinformatics, and molecular approach to determine the causality and relative contributions of environmental, epigenetic, genetic, and stochastic factors to the outcome of disease phenotype is the identification of causal genes and biomarkers with the potential to contribute to complex diseases that have been understudied and/or underreported in relation to a specific phenotype (Fig. 5.6). Integration of multiple genetic and epigenetic variations and environmental and stochastic factors influencing biological pathways and networks will contribute to the understanding of susceptibility to the complex diseases, as well as how they manifest variation in treatment responses. Modern approaches are making strides that need to be included in GEI study designs to pinpoint the cause of complex chronic diseases to identify targets for intervention with a goal to reduce an alarmingly growing public health burden of complex chronic diseases.

5.7   Online Resources

1. Human Genome (http://www.genome.gov/20019523)
2. SNPinfo (http://snpinfo.niehs.nih.gov/)
3. NCBI Database of Genotype and Phenotype (dbGaP) database (http://www.ncbi.nlm.nih.gov/entrez/query.fcgi?db=gap)
4. BioProject (http://www.ncbi.nlm.nih.gov/bioproject)
5. Pubmatrix (http://pubmatrix.grc.nia.nih.gov)
6. Human Gene Compendium's Gene Cards (www.genecards.org)
7. PubMed (www.pubmed.com),
8. The Information Hyperlinked over Proteins (iHOP) Database (www.ihop-net.org)
9. Breast and Prostate Cancer Cohort Consortium (BPC3) http://epi.grants.cancer.gov/BPC3/
10. STRING (http://string-db.org/)
11. KEGG (http://www.genome.jp/kegg/pathway.html)

References

1. Kim, Y.-A., Wuchty, S., and Przytycka, T. M. (2011). Identifying causal genes and dysregulated pathways in complex diseases. *PLOS Comput. Biol.*, **7**(3), p. e1001095.

2. Kunkle, B. W., Yoo, C., and Roy, D. (2013). Reverse engineering of modified genes by Bayesian network analysis defines molecular determinants critical to the development of glioblastoma. *PLOS ONE*, **8**(5), p. e64140.

3. Kunkle, B., Yoo, C., and Roy, D. (2013). Discovering gene-environment interactions in glioblastoma through a comprehensive data integration bioinformatics method. *Neurotoxicology*, **35**, pp. 1–14.

4. Deoraj, A., and Roy, D. (2010). Approaches to identify environmental and epigenomic components or covariates of cancer and disease susceptibility. In *Environmental Factors, Genes, and the Development of Human Cancers*, pp. 197–219, Roy, D., and Dorak, M. T. (eds.) (Springer, New York).

5. Maurano, M. T., Humbert, R., Rynes, E., Thurman, R. E., Haugen, E., Wang, H., Reynolds, A. P., Sandstorm, R., Qu, H., Brody, J., Shafer, A., Neri, F., Lee, K., Kutyavin, T., Stehling-Sun, S., Johnson, A. K., Canfield, T. K., Giste, E., Diegel, M., Bates, D., Hansen, R. S., Neph, S., Sabo, P. J., Heimfeld, S., Raubitschek, A., Ziegler, S., Cotsapas, C., Sotoodehnia, N., Glass, I., Sunyaev, S. R., Kaul, R., and Stamatoyannopoulos, J. A. (2012). Systematic localization of common disease-associated variation in regulatory DNA. *Science*, **337**(6099), pp. 1190–1195.

6. SNP Class Definitions (2005), http://www.ncbi.nlm.nih.gov/books/NBK44488/.

7. Rebbeck, T. R., Ambrosone, C. B., Bell, D. A., Chanock, S. J., Hayes, R. B., Kadlubar, F. F., and Thomas, D. C. (2004). SNPs, haplotypes, and cancer: applications in molecular epidemiology. *Cancer Epidemiol. Biomarkers. Prev.*, **13**(5), pp. 681–687.

8. Turner, S. T., Boerwinkle, E., O'Connell, J. R., Bailey, K. R., Gong, Y., Chapman, A. B., McDonough, C. W., Beitelshees, A. L., Schwartz, G. L., Gums, J. G., Padmanabhan, S., Hiltunen, T. P., Citterio, L., Donner, K. M., Hedner, T., Lanzani, C., Melander, O., Saarela, J., Ripatti, S., Wahlstrand, B., Manunta, P., Kontula, K., Dominiczak, A. F., Cooper-DeHoff, R. M., and Johnson, J. A. (2013). Genomic association analysis of common variants influencing antihypertensive response to hydrochlorothiazide. *Hypertension*, **62**(2), pp. 391–397.

9. Wang, K., and Bucan, M. (2008). Copy number variation detection via high-density SNP genotyping. *Cold Spring Harb. Protoc.*, **2008**, doi:10.1101/pdb.top46.

10. Stankiewicz, P., and Lupski, J. R. (2010). Structural variation in the human genome and its role in disease. *Annu. Rev. Med.*, **61**, pp. 437–455.

11. Erichsen, H. C., and Chanock, S. J. (2004). SNPs in cancer research and treatment. *Br. J. Cancer*, **90**(4), pp. 747–751.

12. Arlt, M. F., Ozdemir, A. C., Birkeland, S. R., Wilson, T. E., and Glover, T. W. (2011). Hydroxyurea induces de novo copy number variants in human cells. *Proc. Natl. Acad. Sci. U. S. A.*, **108**(42), pp. 17360–17365.

13. Carter, N. P. (2007). Methods and strategies for analyzing copy number variation using DNA microarrays. *Nat. Genet.*, **39**(7 Suppl), pp. S16–S21.

14. Feigelson, H. S., Cox, D. G., Cann, H. M., Wacholder, S., Kaaks, R., Henderson, B. E., Albanes, D., Altshuler, D., Berglund, G., Berrino, F., Bingham, S., Buring, J. E., Burtt, N. P., Calle, E. E., Chanock, S. J., Clavel-Chapelon, F., Colditz, G., Diver, W. R., Freedman, M. L., Haiman, C. A., Hankinson, S. E., Hayes, R. B., Hirschhorn, J. N., Hunter, D., Kolonel, L. N., Kraft, P., LeMarchand, L., Linseisen, J., Modi, W., Navarro, C., Peeters, P. H., Pike, M. C., Riboli, E., Setiawan, V. W., Stram, D. O., Thomas, G., Thun, M. J., Tjonneland, A., and Trichopoulos, D. (2006). Haplotype analysis of the HSD17B1 gene and risk of breast cancer: a comprehensive approach to multicenter analyses of prospective cohort studies. *Cancer Res.*, **66**(4), pp. 2468–2475.

15. Thorisson, G. A., Smith, A. V., Krishnan, L., and Stein, L. D. (2005). The International HapMap Project web site. *Genome Res.*, **15**(11), pp. 1592–1593.

16. Harold, D., Abraham, R., Hollingworth, P., Sims, R., Gerrish, A., Hamshere, M. L., Pahwa, J. S., Moskvina, V., Dowzell, K., Williams, A., Jones, N., Thomas, C., Stretton, A., Morgan, A. R., Lovestone, S., Powell, J., Proitsi, P., Lupton, M. K., Brayne, C., Rubinsztein, D. C., Gill, M., Lawlor, B., Lynch, A., Morgan, K., Brown, K. S., Passmore, P. A., Craig, D., McGuinness, B., Todd, S., Holmes, C., Mann, D., Smith, A. D., Love, S., Kehoe, P. G., Hardy, J., Mead, S., Fox, N., Rossor, M., Collinge, J., Maier, W., Jessen, F., Schürmann, B., Heun, R., van den Bussche, H., Heuser, I., Kornhuber, J., Wiltfang, J., Dichgans, M., Frölich, L., Hampel, H., Hüll, M., Rujescu, D., Goate, A. M., Kauwe, J. S., Cruchaga, C., Nowotny, P., Morris, J. C., Mayo, K., Sleegers, K., Bettens, K., Engelborghs, S., De Deyn, P. P., Van Broeckhoven,

C., Livingston, G., Bass, N. J., Gurling, H., McQuillin, A., Gwilliam, R., Deloukas, P., Al-Chalabi, A., Shaw, C. E., Tsolaki, M., Singleton. A. B., Guerreiro, R., Mühleisen, T. W., Nöthen, M. M., Moebus, S., Jöckel, K. H., Klopp, N., Wichmann, H. E., Carrasquillo, M. M., Pankratz, V. S., Younkin, S. G., Holmans, P. A., O'Donovan, M., Owen, M. J., and Williams, J. (2013). Genome-wide association study identifies variants at CLU and PICALM associated with Alzheimer's disease. *Nat. Genet.*, **45**(6), p. 712.

17. Jirtle, R. L. (2009). Epigenome: the program for human health and disease. *Epigenomics*, **1**(1), pp. 13–16.

18. Bernal, A. J., and Jirtle, R. L. (2010). Epigenomic disruption: the effects of early developmental exposures. *Birth Defects Res. Part A Clin. Mol. Teratol.*, **88**(10), pp. 938–944.

19. Lusis, A. J., and Weiss, J. N. (2010). Cardiovascular networks systems-based approaches to cardiovascular disease. *Circulation*, **121**(1), pp. 157–170.

20. Consortium TEP (2012). An integrated encyclopedia of DNA elements in the human genome. *Nature*, **489**(7414), pp. 57–74.

21. Chan, D. C. (2006). Mitochondria: dynamic organelles in disease, aging, and development. *Cell*, **125**(7), pp. 1241–1252.

22. Roy, D., Felty, Q., Narayan, S., and Jayakar, P. (2007). Signature of mitochondria of steroidal hormones-dependent normal and cancer cells: potential molecular targets for cancer therapy. *Front. Biosci.*, **12**, pp. 154–173.

23. Felty, Q., and Roy, D. (2005). Estrogen, mitochondria, and growth of cancer and non-cancer cells. *J. Carcinog.*, **4**(1), p. 1.

24. Hofmann, S., Jaksch, M., Bezold, R., Mertens, S., Aholt, S., Paprotta, A., and Gerbitz, K. D. (1997). Population genetics and disease susceptibility: characterization of central European haplogroups by mtDNA gene mutations, correlation with D loop variants and association with disease. *Hum. Mol. Genet.*, **6**(11), pp. 1835–1846.

25. Booth, S. C., Weljie, A., Turner, R. J. (2013). Computational tools for the secondary analysis of metabolomics experiments. *Comp. Struct. Biotech. J.*, **4**(5), pp. 1–13.

26. Orešič, M., Vidal-Puig, A., and Hänninen, V. (2006). Metabolomic approaches to phenotype characterization and applications to complex diseases. *Expert Rev. Mol. Diagn.*, **6**(4), pp. 575–585.

27. Avery, C. L., He, Q., North, K. E., Ambite, J. L., Boerwinle, E., Fornage, M., Hindorff, L. A., Kooperberg, C., Meigs, J. B., Pankow, J. S., Pendergrass, S. A., Psaty, B. M., Ritchie, M. D., Rotter, J. I., Taylor, K. D., Wilkens, L. R.,

Heiss, G., and Yu Lin, D. (2011). A phenomics-based strategy identifies loci on APOC1, BRAP, and PLCG1 associated with metabolic syndrome phenotype domains. *PLOS Genet.*, **7**(10), p. e1002322.

28. Hunter, D. J. (2005). Gene–environment interactions in human diseases. *Nat. Rev. Genet.*, **6**(4), pp. 287–298.

29. Hahn, L. W., Ritchie, M. D., and Moore, J. H. (2003). Multifactor dimensionality reduction software for detecting gene–gene and gene–environment interactions. *Bioinformatics*, **19**(3), pp. 376–382.

30. Oh, S., Lee, J., Kwon, M-S., Weir, B., Ha, K., and Park, T. (2012). A novel method to identify high order gene-gene interactions in genome-wide association studies: gene-based MDR. *BMC Bioinf.*, **13**(Suppl 9), p. S5.

31. Lage, K., Karlberg, E. O., Størling, Z. M., Olason, P. I., Pedersen, A. G., Rigina, O., Hinsby, A. M., Tümer, Z., Pociot, F., Tommerup, N., Moreau, Y., and Brunak, S. (2007). A human phenome-interactome network of protein complexes implicated in genetic disorders. *Nat. Biotechnol*, **25**(3), pp. 309–316.

32. Charniak, E. (1991). Bayesian networks without tears. *AI Mag.*, **12**(4), p. 50.

33. Jiang, B., Liu, J. S., and Bulyk, M. L. (2013). Bayesian hierarchical model of protein-binding microarray k-mer data reduces noise and identifies transcription factor subclasses and preferred k-mers. *Bioinformatics*, **29**(11), pp. 1390–1398.

34. Oniśko, A., and Druzdzel, M. J. (2013). Impact of precision of Bayesian network parameters on accuracy of medical diagnostic systems. *Artif. Intell. Med.*, **57**(3), pp. 197–206.

35. Glez-Peña, D., Reboiro-Jato, M., Domínguez, R., Gómez-López, G., Pisano, D. G., and Fdez-Riverola, F. (2010). PathJam: a new service for integrating biological pathway information. *J. Integr. Bioinf.*, **7**(1).

36. Druzdzel, M. J. (1999). GeNIe: a development environment for graphical decision-analytic models. *Proc. AMIA Symp.*, p. 1206.

37. Siegel, R., Ward, E., Brawley, O., and Jemal, A. (2011). Cancer statistics, 2011. *CA: Cancer J. Clin.*, **61**(4), pp. 212–236.

38. Zeggini, E., and Ioannidis, J. P. (2009). Meta-analysis in genome-wide association studies. *Pharmacogenomics*, **10**(2), pp. 191–201.

Index